Sheep and wheat are the staples of dryland farms in the Mediterranean zone of the Northern Hemisphere. The commonly used dryland farming system introduced in the 1950s is proving unsustainable. Erosion has reached a critical level and pastures have all but disappeared. Experts advise more cropping (forage crops for instance) and more fertiliser. Yet intensification of the present system will only hasten erosion. Is there an alternative system that is both environmentally sustainable and within the means of most farmers in the region? Innovative farmers in a similar climate in Australia discovered a sustainable rotation using annual medics as both fertiliser and pasture. Attempts to transfer their knowledge have often foundered. Why is this so? How much do the experts know about this system? This book pulls apart the warp and weft of development on dryland farms to try to find some answers to these questions.

T0214712

SUSTAINABLE DRYLAND FARMING

Frontispiece   Farmer instructing technicians in medic farming system in Tunisia at Le Kef.

SUSTAINABLE DRYLAND FARMING

Combining farmer innovation and medic pasture in a
Mediterranean climate

LYNNE CHATTERTON and **BRIAN CHATTERTON**
Independent scholars and consultants

CAMBRIDGE
UNIVERSITY PRESS

CAMBRIDGE UNIVERSITY PRESS
Cambridge, New York, Melbourne, Madrid, Cape Town, Singapore, São Paulo

Cambridge University Press
The Edinburgh Building, Cambridge CB2 2RU, UK

Published in the United States of America by Cambridge University Press, New York

www.cambridge.org
Information on this title: www.cambridge.org/9780521331418

First published 1996
This digitally printed first paperback version 2005

A catalogue record for this publication is available from the British Library

Library of Congress Cataloguing in Publication data

Chatterton, Lynne.
Sustainable dryland farming : combining farmer innovation and
medic pasture in a Mediterranean climate / Lynne Chatterton
and Brian Chatterton.
      p.   cm.
   Includes bibliographical references (p.   ). and index.
   ISBN 0 521 33141 2 (hc)
   1. Medicago – Africa, North.   2. Medicago – Middle East.
3. Dry farming – Africa, North.   4. Dry farming – Middle East.
5. Agriculture – Technology
transfer. I. Chatterton, Brian. II. Title.
SB205.M4C48 1996
633.3′182′096–dc20   95–13514 CIP

ISBN-13 978-0-521-33141-8 hardback
ISBN-10 0-521-33141-2 hardback

ISBN-13 978-0-521-33741-0 paperback
ISBN-10 0-521-33741-0 paperback

The author has provided a website for information. The URL (active at the time
of going to press) is www.drylandfarming.org.

Contents

Preface

This book is about two different systems of dryland farming. One consists of sowing cereals on land that has been ploughed deeply and applying nitrogen and phosphate fertiliser to provide fertility for the crop. Livestock are fed grain and hay and are grazed either in a desultory fashion on poor quality vegetation or not at all. This system is now thought to be responsible for high costs to farmers, high prices for consumers and considerable erosion of farm and rangeland in dryland zones where unreliable rain falls only in the winter time and summers are hot and often windy. The other system is an integrated one in which the land is used for cereal and livestock production in rotation. This system uses naturally regenerating annual legumes to provide nitrogen for cereal crops and pasture for livestock that are grazed all year round. This system requires only shallow cultivation for the cereal crop and none for the regenerating pasture. It is a remarkably low cost system and over time has proved to be capable of sustaining a balance between productivity and environmental stability in difficult dryland conditions.

This book is also about the farmers of South Australia who discovered how to exploit annual legume pasture (*medicago* and *trifolium* spp., commonly known as medic and sub-clover) to solve their dryland farming problems of decreasing production and increasing erosion. It tells of the differences of opinion and approach between them and their technical advisers, who came from institutes where the farming techniques taught were those of the Northern Hemisphere. These technical advisers took the farmers' discoveries lightly and spent many years arguing against them before finally accepting their logic.

The story goes on to examine what happened when development agencies decided to transfer this 'medic farming system' (as it came to be known) to other dryland farmers in the developing world who were losing productivity as their farmland fell victim to erosion. The decision to transfer the knowledge seemed rational enough. In the countries of North Africa and the Near East annual medic and subterranean clover species are indigenous and ubiquitous and the dryland farming zones are sufficiently alike in climatic conditions to those of South Australia to suggest that the integration of medic pastures into the farming system would prove to be the critical factor in solving both the productivity and environmental crises threatening the region.

Because the idea of incorporating medic onto dryland farms came from the remote and relatively unknown Australian farm community the comprehensive knowledge necessary to the successful operation of the system was not always available to the various projects. In addition, the normal problems of technology transfer that tend to arise from the different cultural perceptions of the transferor and transferee were exacerbated by the fact that the indigenous technicians had been educated only in the principles and operations of cereal and livestock production systems used in the Northern Hemisphere. As will be seen, these are often directly in conflict with those of the medic system. Therefore when individual projects finished and the medic farming experts left there was no one for these technicians to turn to within their own institutions to support their brief and often only theoretical knowledge of the new system. As a consequence, the principles of the Australian system were more vulnerable than most outside initiatives to being changed beyond recognition or being lost again.

Surprisingly, given these difficulties, the rationality of the system stood it in good stead and a core of indigenous agricultural technicians and farmers who had welcomed its introduction and been involved in the attempts to incorporate it into their own farming systems continue to try to find means to make it available to their farmers. Their story is contained in last section of this book.

Readers well may wonder how we became interested in the medic system and its transfer. The answer goes back a long time. Brian Chatterton came back to Australia from England after graduating in agricultural science at the University of Reading in the 1960s and took on the task of managing the 400 ha farm 'Riverside' that he inherited in South Australia's Barossa Valley. Originally purchased by his great-great-grandfather (Joseph Barrett) when South Australia was first settled by the British, wheat and sheep and vines were grown on the farm, but, left in the hands of estate managers, it had lost a great deal of its productivity.

'Riverside' is a typical dryland farm in a zone where a winter rainfall (on average 500 mm) provides the moisture for crops and pasture and where summers are so hot and dry that grapes mature full of sugar and make excellent wine. Brian decided to sow the medic and clover pastures being used by many other South Australian dryland farmers to rehabilitate his land and within a few years had achieved levels of productivity never envisaged by his grandfathers. In addition to growing the medic pasture in rotation with cereal crops, he sowed the river banks and the uncleared scrub on the property to annual subterranean clovers and was able to increase the grazing capacity of the property substantially. Once having established the pastures he could say twenty five years later that he rarely had to resow his paddocks – only to extend the pastures elsewhere on the farm.

In 1973 he decided that he could perhaps do something more for farming if he entered politics and by 1975 he had become the State's Minister of Agriculture. He overhauled a cumbersome and, some said, punitive form of drought assistance for farmers to help the rural community through an exceptional drought that was afflicting them. This simplified assistance was designed to support farm families on their farms during the drought and then enable them quickly to sow and re-stock when the rains came. He is credited with having humanised drought finance for farmers.

He then became involved in the State Government's push for overseas markets for its produce and suggested that the dryland farming system common in South Australia should be the basis of a program in which farm technology would be exported in exchange for trade in agricultural manufactures and pasture seed. This was not a new idea – others had it in mind also, notably Dr E.D. Carter, an agronomist at the Waite Institute of the University of Adelaide, and the Western Australian Government's overseas projects unit – but he was in the Cabinet and influential enough to see the plan through to reality.

My rural experience began as a grower of vines, fruit and vegetables in the irrigated Murray Valley of South Australia, but by 1976 I lived in the city and had become an academic and broadcaster. As a result of my studies of rural electorates and my work with the Australian Broadcasting Commission's farmer information program 'The Country Hour' I was asked to join the Minister of Agriculture's staff and later the Premier's, and so Brian and I worked closely together on the dryland farming transfer program.

We planned and held information seminars about the social, political and economic conditions in countries in North Africa and the Near East for technical experts and potential exporters of goods and services; organised the production and publication of books, films and audiovisual kits explaining the medic farming system in diverse languages; and entertained, informed and negotiated with foreign representatives of countries with climates suitable to the system.

In 1979 we were sent on an eight week tour of the Near East and North Africa on behalf of the South Australian Government – negotiations being undertaken by Brian at the highest official levels and inspections made of ordinary farms and research institutes in eight countries.

The vicissitudes of politics put us both into Opposition and during this time we returned to North Africa and the Near East once or twice every year for extended periods using our own resources, but with the full cooperation of Ministries of Agriculture. We conducted deeper studies of the region's farms and farmers and the rangeland and its livestock owners. We were given *carte blanche* to go anywhere we liked and were always accompanied by a technician who began by translating for us and usually ended by becoming an advocate of the use of medic pasture on dryland farms and an expert in explaining techniques necessary for its success to farmers and others.

Without exception at the end of these tours we were summoned by Ministers and senior officials to give advice on the manner in which they should plan for improvements to the existing systems. On many occasions this advice was accepted and action undertaken to put it into operation.

By 1983 we had begun to publish papers on various aspects of farming and rangeland exploitation in North Africa and the Near East and when the government in South Australia changed again, Brian was re-elected Minister of Agriculture. Unhappily, the projects established with such hope at the end of 1979 had not gone well due to a number of factors, some political, some operational. It was not possible from within government to repair the damage caused and renovate the project program, and Brian resigned.

In reflecting over our experience, we were troubled by the fact that so few decision-makers in the relevant institutions and agencies knew about, let alone understood, the principles of the medic farming system and even fewer realised what was needed at the farm level in order to enable farmers to try it out properly. We suspected, for example, that the reason why appropriate equipment was not ordered by the agencies when medic projects were mooted was simply because they did not realise that it was necessary. They seemed to believe that medic was yet another forage crop and that while it was interesting it simply slotted in to the existing system and required no fundamental change to the conventional shopping list for a dryland farming project.

We decided to begin an intensive effort to make the medic farming system more widely known throughout the dryland farming world and within the large funding agencies that controlled the development process in client countries. We continued to visit North Africa and the Near East regularly and tried to create a network of encouragement for those working on dryland farming projects using medic pastures. In 1986 we managed to persuade the Food and Agricultural Organisation in Rome to pay for the production of four audiovisual kits in the language of each country which we filmed on small farms in North Africa and the Near East to show technicians and farmers how to establish and deal with the operations of growing cereals and grazing livestock using medic pastures. We were employed as consultants by the United Nations and by Arab organisations to review cereal and rangeland projects and to draw up national plans for training and demonstration programs.

During these exercises, we learnt many things about the paucity of information available to local technicians, and how they suffer due to an almost non-existent distribution network for what there is. We were able to persuade the FAO to provide funds to produce a manual on the use of medic pastures in North Africa and the Near East to assist technicians and extension agents who needed data and documentation to enable them to understand the way in which the system worked. This taught us a lot about the pitfalls of publishing information through large international agencies.

We found Australia too far away to be able to make regular study tours or to undertake useful consultancies or attend conferences relevant to North Africa and the Near East, and so in 1986 we moved to Italy where we now live. For a time we sent out twice yearly a medic newsletter to try to keep those working on medic in touch with latest developments, and we were encouraged by the support we got from Australian medic farmers and the technicians and decision makers who received it.

In the course of these attempts to widen the extent of knowledge about the medic farming system we have been struck by the distance that separates the many useful books written and published on the global effects of agricultural development in dryland farming regions, the scientific publications of specific data relating to these regions, and the reports of those who have worked on projects designed to improve the farming there.

The former two categories are available widely and form the basis of many development studies programs and are included in course work in agricultural science. The latter are seldom read by more than the writers and the administrator who receives them.

Although there is considerable intellectual criticism of global aspects of development in the Third World countries and often individual projects are held up to scorn because of their insensitivity and ineptness, there has been little attempt to critically analyse the delivery of information and the performance of competing dryland farming systems from the farmer's point of view.

The medic farming system is fundamentally different from the dryland farming system now being practised in most countries around the Mediterranean basin and so provides a focus for such an analysis. The major principles of its operation were first discovered on farms in South Australia and it became established in isolation from the then prevailing scientific engine that was driving agriculture in the more humid cereal zones of the Northern Hemisphere in another direction. Farmers' records allow us to understand why the differences occurred and why Australian farmers decided not to discard their own discoveries and conform to the European and American consensus on dryland farming.

Can we say that one system or the other is better, and can we say that one group of transferors are more effective than another? This takes us into a particularly sensitive aspect of technology transfer and agricultural development.

Are Western agencies, heavily influenced by conventional European attitudes to dryland farming, always right in depending solely on technical experts in order to carry out farm improvement programs?

Should those who believe that an expert farmer is better at transferring technology at the farm level be dismissed as 'romantics'?

The recounting of the origins of the medic farming system and its travels to North Africa and the Near East provides an opportunity to show that farmers are capable of finding intelligent and scientifically sound answers to environmental and farming crises. The simple and practical solutions invented by farmers when faced with a farming problem often make it unnecessary to wait upon years of research and the sifting through of data that is the conventional technical response to farming problems. Farmers have a different perspective about farming to that of the scientist and technician. There has been a trend in recent times towards a re-evaluation of the wisdom of the farmer in meeting the challenges of improving a farming system. It has not resulted in the big agencies or institutions giving more attention to that wisdom. The role played by South Australian farmers in the evolution of their dryland farming system and the role they have played in transferring its operations to other farmers beset with the problems for which it was developed provides, we believe, a case study of farming expertise in action.

This brings into prominence the further question of whether aid for agricultural development should remain the fiefdom of scientific and technical experts and the research institutions alone, or whether development projects need to put more emphasis on employing those who are capable of making effective changes at the farm level. If this is so, the expert farmer may well become the engine driver of change rather than the stoker.

This has been a large task to attempt. No doubt we have left a great deal undone and probably not succeeded terribly well in our purpose. We hope, however, that this imperfect work will spark off more critical thought and discussion about the resources urgently needed on farms in the dryland farming zones of the Northern Hemisphere so that the crisis of erosion and degradation already destroying the soil on these farms will eventually be halted.

Our sources for this work are comprehensive. We have been extremely fortunate to have many friends and colleagues who have made their own important contributions both to the development and to the transfer of the medic system, and they have been wonderfully generous with their contributions to this book. During our studies we have accumulated a large amount of material relating to the cultivation and use of medic pastures throughout the world. A lot of the material is in the form of unpublished papers prepared by others as well as ourselves; interviews we have conducted with farmers, technicians, scientists and administrators, some of which has been transcribed, some of which remains on audiotape or in note form; and many unpublished intergovernmental reports, memos and other documents that refer to project negotiation and the planning and operation of various projects. We both take extensive notes during study tours and consultancies and either one of us or both have kept regular diaries. Copies of diaries, various reports and some documents have been lodged with the Barr Smith Library of the University of Adelaide. The originals of all material used for this book are at present in our own collection, which is referred to in the references as 'Chatterton Papers'.

We did also have a unique collection of photographic slides that recorded medic pastures sequentially over a decade or more in various parts of the dryland world. Unfortunately a massive burglary of our home in Australia threw our collection into disarray and although we have been able to restore some order to the written and tape recorded material, it has been impossible to do this with the photographic record. We are always happy to make copies of relevant material available to other scholars.

We have been fortunate to have the patient editorial assistance and encouragement of Dr Alan Crowden and Dr Maria Murphy of Cambridge University Press and, in the initial stages, that of Dr Robin Pellew. Dr Tony

Allan and Dr Tony Griffiths have been kind enough to read the manuscript at various times and made helpful suggestions. Dr Tony Allan has, in addition, given us many hours of his valuable time, often at very short notice, to comment critically on many chapters and to advise on ways in which sections of the manuscript could be better organised. Although I have written the text of this book the experience and the research on which it is based has been undertaken in partnership with Brian Chatterton, without whom this book would not have been possible. The errors are mine.

05010 Montegabbione (TR) ITALY   *Lynne Chatterton*

Part one

Medic and other systems

1
Why use medic?

Introduction

Dryland farmers in North Africa and the Near East have, in the past fifty years, been the focus of many schemes to try to improve the productivity of their farms. Many of them are battling to maintain levels of productivity due to erosion and decreasing soil fertility. The system they use consists of various versions of a cereal/fallow cycle. In latter years it has been suggested that they replace the fallow with grain legumes or forage crops, and use more nitrogen fertiliser. To a large extent farmers have resisted this advice. An alternative system is available that is cheap and particularly suited to dryland farms where capital is scarce and resources are few. This is the medic/cereal rotation that was discovered by farmers in South Australia early in the twentieth century.

The legume pastures used as the basis of livestock and cereal production on dryland farms in South Australia consist of two major species.

Medic

Medic is the common name given to species of annual medicago (*M. truncatula*, *M. rugosa*, *M. polymorpha* etc.) that are native to the winter rainfall zones of North Africa and the Near East, the Southern Mediterranean zone of Europe and Turkey. They are not native to Australia but they adapted remarkably well to Australian conditions. Several medics are believed to have been introduced accidentally onto farmland in South Australia in the mid-nineteenth century as contaminants of imported cereal seed from North Africa. Others have been introduced later as a result of programs of selection for particular ecological niches. Medics in their natural habitat flourish in the dryland zones where average annual rainfall is between 100 and 550 mm and soil is alkaline. They exist in varieties that

3

tolerate drought and cold and relatively high altitude. They evade drought. There is a close relationship between their normal period of growth and that during which moisture is available in the surface soil. Medic species produce a fairly high proportion of hard (i.e. dormant) seed and this enables them to tide over a season in which (due to drought) little or no seed production takes place. Subsequent rain will germinate the dormant seed and re-establish the stand of pasture (Trumble, Whyte & Nilsson-Leissner, 1953, p. 211).

Medicago sativa

Annual medic must not be confused with *Medicago sativa*. *M. sativa* is commonly known as lucerne or alfalfa. *M. sativa* is a perennial plant which requires periodical re-sowing and does not survive adequately if a cereal crop is sown onto land on which it is growing. The drought resistance of lucerne is not because it needs little water; it is 'one of the most wasteful forage plants as far as water is concerned' needing 750 tons of water to build up 1 ton of cured hay. Unlike medic, which uses the immediate rainfall, the tap root of the lucerne plant draws up ground water to satisfy its habit (Trumble *et al.*, 1953, p. 211).

Sub-clovers

Annual clover species *Trifolium subterraneum* and *T. yanninicum* are similar in their growth and performance to the annual medic species. They are commonly known as 'sub-clover' and are used in the same way as medic but on acid rather than alkaline soil. *T. brachycalcinum*, another variety of sub-clover, is suitable for neutral to alkaline soils. Sub-clover will tolerate low average annual rainfall (300 mm). Generally, sub-clover does not have the same quantity of dormant seed as medic and for this reason needs re-sowing if grown in rotation with cereals. There is some doubt about this generality. On our own farm, where the soil is on the cusp of acidity and where medic pastures and sub-clover have been grown in rotation with cereals for twenty five years as well as being used for permanent pasture, sub-clover (cultivars Clare and Dwalgenup) has survived all farming operations and regenerated as well as the medic and has not been re-sown during this period. Sub-clover pasture was first recognised in the latter part of the nineteenth century when a South Australian grazier, faced with declining natural pasture on his property, encouraged the growth of naturally occurring sub-clover and later harvested seed to use on other pasture land and found that it provided a sustainable basis to his livestock enterprise. Permanent sub-clover pasture is now the basis of dairy and beef cattle production in hill country where rainfall is in the vicinity of 500–550 mm and also for flocks of sheep in drier areas with acid soil.

The logic of medic farming

The medic farming system exploits the plant in the following manner. The green medic plant provides a high quality green pasture that grows best when constantly grazed throughout the winter–spring growing period. Medics produce pods in spring that contain seed, a large proportion of which have a hard coat that enables the seed to remain dormant for up to three years. After senescence in the summer the resulting straw and the dry pods and the seed within the pods provide abundant, high protein feed for livestock. Some of these pods need to be safeguarded from grazing so that sufficient seed is left in the ground to regenerate a good pasture following subsequent autumn rains. If the rain comes out of season, a proportion of the seed will sprout but there will remain sufficient in a dormant state to provide the source of new plants when the appropriate seasonal rain does fall (see Figure 1.1).

When managed correctly (the number of sheep grazing are determined by the amount of pasture available) the pasture can supply the bulk of feed needed by livestock all year round. Medic and sub-clover pastures provide this remarkably cheap and abundant source of high quality livestock feed because they do not require annual re-seeding if the management is adequate and, because they thrive under grazing conditions, they relieve the farmer of the need to make and feed out large quantities of hay, or to buy in expensive quantities of concentrates. They do require an annual dressing of phosphate to encourage abundant growth. Current annual usage of superphosphate on South Australian dryland farms is about 150 000 tonnes (100 kg/ha on average) for wheat, 150 000 tonnes for other grain crops and 200 000 tonnes for pasture (SAYB, 1985, p. 370).

The farmer does not need to grow cereals in order to perpetuate the medic pasture but if he wishes to have an integrated system, the benefit of correctly managed grazing carries over into the cereal phase of his rotation. Medics are leguminous and fix nitrogen from the air in the soil so that crops grown after medic benefit from this. The farmer need not purchase nitrogen fertiliser for his crop. Webber and his colleagues record that measurements taken after the first year of medic in the cereal zone of South Australia

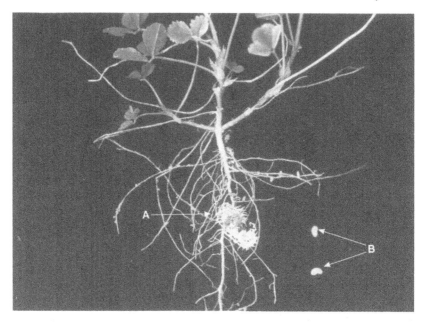

Figure 1.1. A medic seedling removed from the soil in early winter. The seedling has grown from a seed in the pod (A). The pod was broken open to reveal two more seeds (B). These did not germinate, but will do so in future seasons. (Photo: F. Botts.)

showed an average increase of 70 kg/ha of N and that an increase of 200 kg/ha of N had been recorded on sandy soil after 'a vigorous sward' of *M. littoralis* (Webber, Cocks & Jefferies, 1976, p. 32). The annual net increment in soil nitrogen from grazed medic and sub-clover pasture averages around 100 kg N/ha (Carter, 1975, pp. 1–20). The future quality of the regenerating pasture during the cereal phase depends on the safeguarding of the dormant medic seed that remains in the soil. This is achieved by using shallow tillage to a depth of not more than 10 cm during the cultivation and seed bed preparation program. If, during the cereal phase, the land is ploughed deeper than this most of the medic seed will be buried too deep to regenerate. Repeated ploughing will deplete the seed bank until the re-seeding of the pasture will be necessary. This will be costly and will intensify the cultivation of fragile soils and nullify many of the benefits of the system. In addition to safeguarding the medic seed the farmer gains significant reductions in costs and energy required to grow a cereal crop. The operations of shallow cultivation, weed control and seed bed preparation are carried out using a robust scarifier with appropriate points. The program requires no more than three passages over the soil and takes five weeks compared to the five months common with the European-style cropping system because deep ploughing needs subsequent frequent passages in order to break down clods caused by the initial ploughing. The farmer who adopts a medic/cereal rotation can benefit further from the use of a combine seeder. This machine, incorporating a scarifier, seed and fertiliser boxes and harrows, carries out the operations of seed bed preparation, placing of seed and phosphate fertiliser and light cover of both in one passage over the ground. Because it is easy to change the mix of cereal and livestock production each year in response to both climatic variations and market signals, the system allows the dryland farmer flexibility in his operations and enables him to grasp the opportunity of good seasons while not suffering unduly from the bad. The medic system is a cheap option (much cheaper than that dependent upon deep ploughing, purchased fertiliser and manufactured feed) and is within the resources of the capital poor, dryland farmer.

The effects of a medic farming system are

- better protection from erosion
- less frequent cultivation of the soil
- the build up of organic matter during and following grazing
- an enhanced ability of the soil to absorb rainfall more effectively.

These attributes of a medic farming system enable it to be used in those marginal zones where average annual rainfall is below 250 mm, where it has long been believed that cereal production and grazing cannot be undertaken without creating and intensifying erosion.

It is, however, essential to realise that growing a particular cultivar of medic or sub-clover on its own cannot work miracles. An understanding of how and why the various aspects of the system interconnect and work is the key to successful management. It is this management that enables the farmer to manipulate the system to his advantage and to evade the worst consequences of drought or market down turn.

This discovery that medics and sub-clover pastures could be the basis of a sustainable farming system came about in the early part of the twentieth century after farmers in South Australia incorporated spontaneously occurring medics into their existing but failing farming system and found fairly quickly that their productivity rose and the erosion and infertility of the soil was reversed. The twentieth century has provided a record in South Australia of sustainability and productivity on farms and pastoral properties where medic or sub-clover, or both, have been the basis of the farming system. For the rest of the world the discovery of the exploitation of medic and sub-clovers in a deliberately manipulated farming system to provide sustainable cereal and livestock production is little known, yet it promises to help restore sustainable farming to many of the dry zones with winter rainfall currently being threatened by environmental degradation. Attempts to transfer the knowledge gained in Australia to dryland farmers in North Africa and the Near East have already provided sufficient evidence to lead us to suppose that medic and sub-clover pastures incorporated into a suitable farming system would improve the fertility of the soil in the cereal zone and provide a substantial increase in the availability of livestock feed enabling some pressure to be taken off the marginal and rangeland zones that could then, in their turn, be sown to permanent medic pasture and suitable fodder shrubs. The revegetation of this zone has long been the objective of those concerned at the continuing degradation and desertification that has accompanied unwise farming and uncontrolled grazing. The effect of medic pasture on soil fertility is obvious on farms where it has been used after a number of years. Dry, caked soil that resists rainfall or that dissolves into dust when clenched in the fist has little fertility. Soil that clings, that is friable, that has visible organic matter, and that absorbs rainfall evenly is fertile. The two illustrations here show the difference in soil where bare fallow is practised (Figure 1.2) and soil where medic/cereal has been the rotation for a number of years (Figure 1.3). The measurement of increased productivity from better structured soil, a greater quantity of organic matter, and soil nitrogen that follows the use of grazed medic pasture is contained in the data of individual farms, project sites, and research centre trials that are detailed in the following chapters.

Why dryland and not rainfed?

The term 'dryland farming' is used in this book to describe farming where the growing season is winter and early spring. This season is triggered by late autumn/early winter rains. There is little or no rain in summer. Cereals are grown in the zone with an average rainfall ranging from 200 to 500 mm. In any given year the variation from the average can be considerable. Below an average of 200 mm the zone is usually categorised as rangeland. The boundary is not precise and in the range of 150–250 mm of average rainfall there is a zone that we have categorised as marginal. The zone above 500 mm average rainfall is often hill and mountain country and has more complex mixes of livestock, cereals and horticulture. One sometimes comes across the term 'rainfed' to describe farming in these zones, but this can be

Figure 1.2. Photograph showing caked soil after continuous cereal/fallow farming, without medic.

Figure 1.3. A handful of soil, sheep droppings, straw and medic pods after rotation.

confusing. 'Rainfed' agriculture is also applied to farming without irrigation in the temperate zone, for example the cereal farms of France and England. 'Rainfed' agriculture can take place also in sub-tropical and tropical zones. The dry tropics of Africa and India are examples. Australia in the north – Queensland and the Northern Territory – has sub-tropical 'rainfed' agriculture, but in this zone there is summer rainfall and the winter is comparatively dry. The types of pasture plants and breeds of livestock best suited to these zones are different to those suited to the dryland zone with a Mediterranean climate.

There is a huge difference in the conditions under which cereal is grown in England compared to the conditions under which cereal is grown in countries with Mediterranean climates, although the annual average rainfall may be the same. Similarly, different types of pasture plants and breeds of livestock are used.

A further reason for retaining the term 'dryland farming' is that it has become current in North Africa and the Near East where the transfer is being carried out and has become synonymous in many parts of the world with the Australian farming system based on medic pastures. It would be a pity to upset this simple connection.

Why dryland farming instead of irrigation?

In North Africa and the Near East, in spite of some grandiose potential schemes and large amounts of capital invested in the Great Man Made River in Libya, the plans to tap the Albion Nap in Algeria, and dams in Iraq, Morocco and Jordan, irrigation has failed to provide the promised results of vastly increased production of cereals and livestock. In Kuwait and Saudi Arabia and the deserts of Libya, where we have seen irrigated wheat and lucerne grown with sophisticated sprinkler systems and chemical fertilisation, the results were spectacular to begin with. After several years, the initial fertility has been rapidly used up and the economic viability of the schemes has been undermined, not only by the capital cost of exploiting the water, but also by the need to constantly re-invigorate the soil, and the cost of transport of the produce to far-off markets. Dr J.A. Allan has argued that the use of previously untapped and remote supplies of renewable and non-renewable water for cereal and livestock production is not a sound economic use of scarce water. 'Water allocated to agriculture in hot countries is always an expensive, if not an extravagant, option' (Allan, 1989, p. 123). The assessment of irrigation potential in the region is that a very small proportion of the land can be irrigated. Given that even if at some future date viable schemes for the use of this water for agriculture (i.e. cereal and livestock production) could be brought to fruition, it will hardly provide an economically viable solution to the growing demand for mutton and grain that we see today in the region. Even if it were economically viable, it is impossible to distribute irrigation water to thousands of small dryland farms, so one is faced with the question of what to do with them. If nothing is done the probability is that the farmland will continue to erode and families will be forced to abandon it. There is a need to improve the productivity and profitability of these farms for the sake of the families, if only to try and prevent further migration into the already overcrowded cities. If irrigation is not the answer to the deficits, then the question arises, can the dryland cereal zone and the rangeland be improved? If the use of medic pastures can be established as the basis of cereal and livestock production in the region, it is likely that the problems of producing sufficient grain and mutton to supply the growing demand by consumers can be met without the need for more doubtful investment in irrigation (Carter, 1975).

In South Australia, where the ratio of dryland to that of land available for irrigation is similar to that of most of North Africa and the Near East, the use of irrigation for cereal and livestock production has not been an important factor. Capital investment in irrigation from the only reliable river (the River Murray) has been made for the most part to support Soldier Service Settlements on small holdings to grow vines, fruit and vegetables after the two World Wars. It has not been an economically productive strategy. In the first instance the government was left with a debt of £3 million ($AUS 6 million), and in the other the settlers suffered from accumulated debt as well. There is some irrigation of lucerne (alfalfa) for dairy cows on the lower swamps of the river, but apart from those and a few individual blocks for irrigated lucerne using underground water, the only other irrigation has been in the South East and the plain north of Adelaide where ground water is used for irrigated seed production and some lucerne. On the whole, irrigation is a negligible factor in the production of cereals and livestock (Williams, 1974, pp. 227–60).

Economic investment in dryland farms, once the land has been bought or allocated, consists mainly in machinery, fertiliser and seed used by the cereal farmer and, in the case of livestock production, the cost of the initial flock and the continuing cost of feed. If the cost of these investments can be kept relatively low and a reasonable level of productivity of the farm or animal sustained, then dryland farming is both attractive to the farmer and economically sound.

(a)

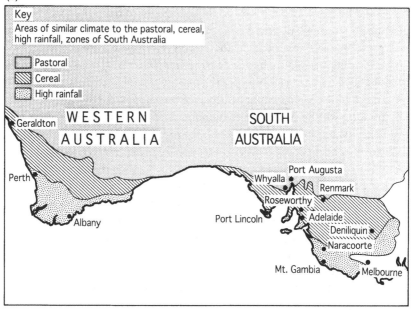

(b)

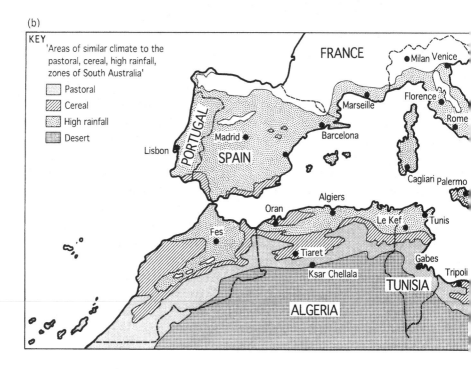

Comparisons between Southern Australia and the Near East and North Africa

Climate

The climatic similarities between South Australia and Near East and North Africa where dryland farming takes place are illustrated in the maps shown in Figure 1.4.

Productivity increases due to annual legumes

Productivity increases on dryland farms and the reasons for them are difficult to quantify, not the least because of the swings in seasonal conditions and the large number of factors that affect productivity on individual farms. Increases are easy to measure on plot experiments but plot experiments have little validity when applied to whole farms or regions. We have drawn a number of graphs below that illustrate the different levels of productivity achieved over a relatively long term on farms

Figure 1.4. Maps showing comparison of climatic and farming zones of Western and South Australia (a) and the Near East and North Africa (b), compiled from data in Bioclimatic Map of the Mediterranean Zone (UNESCO-FAO, 1963, pp. 23–5), courtesy South Australia Department of Agriculture.

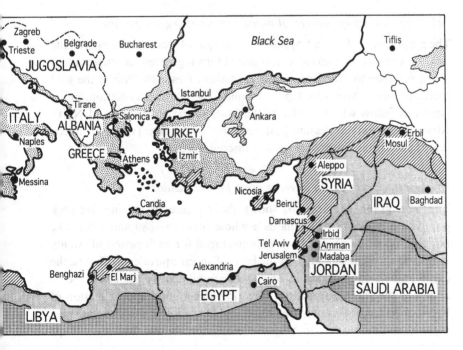

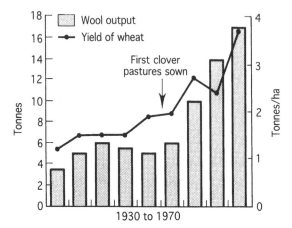

Figure 1.5. Wool output and wheat yields at Turretfield Research Centre, 1930–70. Reproduced from Webber *et al.* (1976) with permission.

in South Australia where medic and/or sub-clover pastures provide year round grazing for livestock and fertility for cereal crops and on farms in the Near East and North Africa where fallow, nitrogen fertiliser and conserved fodder and grain are used.

(a) Integrated sheep and cereal production using legume pasture

The graph above (Figure 1.5) shows the output of wool (columns) and the yield of wheat (thin line) for an integrated farming system of livestock and cereal production practised on the Turretfield Research farm in the Mid North of South Australia. Up to the mid-1950s the main rotation on the farm was fallow/wheat similar to that used in the Near East and North Africa. After that time legume pastures (mostly sub-clover) were introduced into the rotation and fallowing was discontinued.

(b) Livestock population increase using legume pasture

The next graph (Figure 1.6) shows the increase in grazing livestock population in South Australia as a whole between 1880 and 1980. The populations in thousands have been averaged for each period of twenty years and then consolidated on the basis of sheep equivalents. That is, the cattle and horse populations have been multiplied by ten before being added to the sheep population. This provides a good approximation of livestock trends but is not a conversion into Dry Sheep Equivalent (DSE), a measure commonly used in Australia and one that would require each type

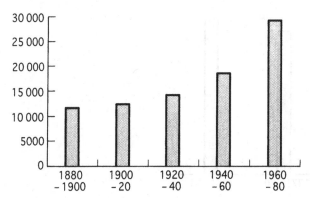

Figure 1.6. Livestock populations in South Australia, 1880–1980. Source: SAYB (1970, 1983).

of livestock to be subdivided into separate classes and multiplied by a different factor. We have done this because the grazing of pasture has moved back and forth from horses and sheep to cattle at various periods due to changing price signals. For example, in the 1970s there was a beef boom and cattle numbers rose to a peak of 1 891 000 in 1975 from a base of 690 000 in 1965. By 1984 when prices dropped the number went down to 800 000 (SAYB, 1985, p. 595). The population of horses was quite significant between 1911 and 1924 when they were used for traction and transport and amounted to more than 2 500 000 sheep equivalents.

The substantial increase in the numbers of livestock grazing pastures during the period 1940–80 reflects the use of medic and sub-clover on farms both as permanent pasture and in rotation with cereals. However, other innovations account for some of the increase and should not be overlooked. For instance, during this period the discovery of trace elements enabled an expansion of pastures into areas such as the upper south east of the State and parts of Kangaroo Island where formerly it was not considered worthwhile to put in fences and watering points (Williams, 1974, pp. 321–4). Simultaneously there has been an increase in wool and meat production per head due to better nutrition from a constant diet of good quality pasture and dry straw and pods. The effect of better breeding and animal health programs will account for part of this but the high quality feed from annual legume pasture is considered to be the major cause. Australia-wide the population of grazing animals (in this case, sheep and cattle) has increased from 157 million sheep equivalents in 1902 to 357 million in 1983 (YBA, 1984, p. 285). There is no evidence to suggest that

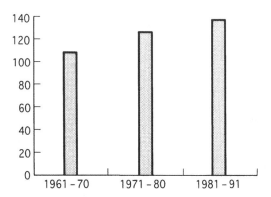

Figure 1.7. Livestock populations (shown in millions of sheep equivalents) in the Near East and North Africa, 1961–91, from an FAO Agrostat database printout (21.12.92).

population or productivity gains have come from feeding grain or other fodder.

Figure 1.7 shows the population of sheep and cattle converted to sheep equivalents on the basis of multiplying the cattle numbers by ten and adding the sheep number, for the countries Morocco, Algeria, Tunisia, Libya, Jordan, Syria and Iraq, which have been selected as having similar dryland zones to Southern Australia.

It can be seen that there has been an increase of about 29 000 000 sheep equivalents or 26% during the thirty year period. This broad increase breaks down into a strikingly uneven pattern of livestock population growth as shown below.

This next graph (Figure 1.8) shows changes in livestock populations in each country. The negative figure for Iraq may represent a decline in livestock numbers due to the Iraq–Iran war rather than a decline in available nutrition. The other countries can then be put into two groups. The first group consists of Tunisia (plus 4.2%), Jordan (plus 15%) and Morocco (minus 2%) where livestock populations have not moved very much over the last thirty years. The relatively larger increase for Jordan does not seem outside the normal range of population growth over time for dryland zones where droughts and good seasons cause significant fluctuations. In the other group we have Syria (plus 109%), Libya (plus 139%) and Algeria (plus 135%) all of which show a very large increase. Algeria's increase accounts for more than half that for the region. The low figure from which this increase is measured (which is only two million more than Tunisia – a very much smaller and less fertile country) may be due to the

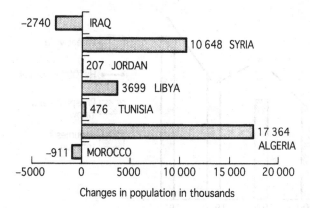

Figure 1.8. Changes in livestock populations in the Near East and North Africa, 1961–91, from an FAO Agrostat database printout (21.12.92).

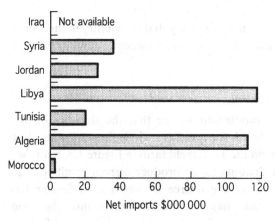

Figure 1.9. Cost of net imports of feedstuffs for countries in the Near East and North Africa, 1985–90, from FAO Yearbook (1990).

vicissitudes of the Algerian war of independence that did not end until 1962. It is difficult to explain the increase in Libya and Syria.

(c) Productivity and imported feedstuffs

The most obvious explanation for such increases is the amount of imported feedstuffs, and Figure 1.9 shows the average annual cost of net feedstuff imports for the countries shown above for the period 1985–90 in $US.

The feedstuff bills do not appear to relate directly to the population

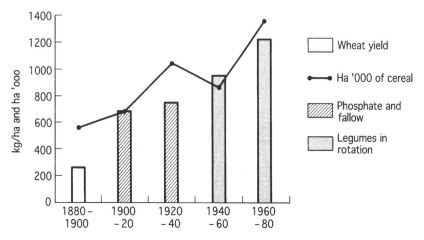

Figure 1.10.   Wheat yields and cereal area for South Australia, 1880–1980.
Sources: Donald (1964) and SAYB (1985).

increases indicated above, but it is fair to say that no country in the region
has achieved a large population increase in its livestock without incurring a
large bill.

(d) Productivity of cereals in rotation with medics

If we turn now to cereal production we see that the statistics on the
increase in cereal yields for South Australia as a whole show a less startling
increase than those shown on the Turretfield farm in Figure 1.5, but it has
to be remembered that individual farms produce uneven results due to
management regimes reflecting the degree to which each farmer has
adopted legume pastures and has incorporated them into the farm
program. Figure 1.10 shows wheat yields (columns) and the area of wheat
and barley (line) in South Australia since the 1880s and is based on a
similar one produced by C.M. Donald but with the addition of areas sown
and is extended to 1980.

One can assume that the introduction of superphosphate between 1900
and 1920 was responsible for the leap in productivity shown during that
period, because the area under crop did not change much. A much larger
increase in area took place between 1920 and 1940. It was during this period
that there was an expansion out into the drier areas, but it was also during
this period that the use of bare fallow before wheat reached its peak. The
increase in yields is not great. Between 1940 and 1960 the area under crop
decreased a little but yields showed a marked increase and this reflected the
abandonment of bare fallow and its replacement with legume pastures in

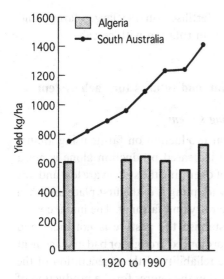

Figure 1.11. Wheat yields in Algeria and South Australia, 1920–90. From FAO Yearbook (1991), Perkins (1927) and SAYB (1985).

the cereal zone. Between 1960 and 1980 legume pastures were established on most farms to a greater or lesser degree and the extension of the area under crop has taken place together with an increase in yield. While the major innovations of superphosphate and legume pastures account for most of the increase, plant breeding and the adoption of herbicides for weed control must also be given credit. Statistical increases in the area under crop can reflect either the extension of cereal growing into more marginal areas with inherently lower yield potential, or more intensive rotations in existing cereal growing areas. In the case of South Australian dryland farms the gradual changeover from bare fallow to legume pasture and superphosphate has resulted in a fourfold increase in cereal yields over one hundred years and a virtual doubling in the past fifty.

Cereal productivity without medic pasture

Figure 1.11 compares the trend in wheat yields in South Australia with that in Algeria during the period 1920–90.

This illustrates the consistent increase in yield in South Australia where legume pastures are used and the stagnation of yield in Algeria where the yield has hardly increased during the period under review. This is in spite of the mechanisation of many operations on farms in Algeria since the 1950s, the use of nitrogen and phosphate fertiliser on the larger farms, and attempts to introduce forage legumes in rotation with cereals.

Comparisons of costs of production and returns for each system

Costs associated with a medic farming system

The increases in livestock and cereal production on farms using annual legume pastures are significant, but increased production alone is not a justification for their use. The costs of operating the system are low and this has been a strong motive in farmers adopting it in the first place and then making it the permanent base of their dryland farming. The initial cost of the medic or sub-clover seed to establish the pasture is not great and resowing is only necessary following prolonged drought or bad management of the grazing. An indication of the reliability of the regeneration of the pasture in South Australia is reflected in the figures for the production of medic and sub-clover seed. In 1984, 2300 tonnes of sub-clover seed was produced and just over 2400 tonnes of medic seed (enough to sow about 200 000 ha of pasture), a large proportion of which was exported to other states in Australia and overseas. About three million ha of medic pasture are being grazed on farmland in South Australia in any given year, (SAYB, 1985, p. 378, Seedco, 1988). The other major cost is an annual application of phosphate. As part of the farm each year can be in cereals and another part in medic, the availability of cereal stubble and medic pasture releases the South Australian livestock owner from the need to buy in other livestock feed except when there is prolonged drought. Even then, if he has been prudent, he will have made a hay reserve from dry medic in years of abundance, and also put aside cereal straw for such an eventuality. A major cost benefit to the dryland farmer of a medic/cereal rotation becomes evident after a period of drought. During drought the dryland farmer will not harvest a cereal crop and will frequently lose the seed and fertiliser he has used to plant one. His income from livestock will also be down and he will have incurred costs of buying in feed to replace the lack of home-grown hay or forage. His cash reserves will be non-existent or low. With a medic/cereal system the medic pasture will regenerate when the first rain falls in the normal season with no cost to the farmer, and he can rapidly utilise it for grazing. He may be able to buy more sheep to restore his flock numbers, but he can decide on the amount of pasture to utilise. If he cannot afford to buy more sheep, he may find it useful to rent out some of the pasture, or, if it is well established he can cut the excess for hay for store or sale. The rapid and shallow cultivation used for the cereal phase costs much less than deep ploughing. The availability of nitrogen from the pasture for the cereal crop is not affected by the drought period and the farmer need not buy nitrogen fertiliser. He must buy only seed and superphosphate.

Costs of wheat/bare fallow system

The cost of cereal production in the Near East and North Africa is very high. One cereal crop grown on bare fallow involves the land for two years to grow one crop and must include the cost of cultivating that fallow, the high cost of deep ploughing, the frequent passages of the soil needed to break down the large clods left after the ploughing and the subsequent separate passages during which the farmer applies nitrogen and phosphate fertiliser and sows seed. Studies carried out in Algeria, Tunisia, Morocco, Jordan, Libya and Syria indicate that if the farmers of the region adopted the cheap and efficient tillage required by the cereal phase of the medic farming system their costs of production would fall significantly. As an example, in Algeria the existing seedbed preparation program (including application of phosphate fertiliser) for cereals costs 112 AD/ha. When the farmer uses the medic farming tillage program his costs fall to 42 AD/ha (Chatterton & Chatterton, 1987–90). Farm-grown hay and forage for livestock involves soil preparation, seeding and fertiliser, together with the cost of cutting and later feeding out. Costs of re-establishment in the region after a drought are correspondingly high and cause a strain on the farmer's cash reserves or force him to borrow in order to continue farming.

Returns from livestock in the medic system

The South Australian farmer receives a relatively low price for his lambs (rarely does the price go much over US$15 per 30–35 kg/animal) and well grown wethers fetch little more than US$20 for export to the Arabian Gulf, and often less at the market. Farmers in South Australia would expect to produce around 2000 kg of meat per year from 100 ewes. The price of wool is a greater incentive to production than that of meat and until recently a price support scheme has made this a stable outlet for the livestock owner. A well grown Merino will produce about 6 kg wool per year. Australian wool is highly sought on international markets and a price of around US$6/kg (clean) was being obtained in 1990. In late 1990 the demand for Australian wool was deeply depressed due to oversupply, and in November of that year the Australian Government was considering a scheme to destroy about 40 million sheep out of the total national flock of 120 million owing to a lack of market for sheep products. In February 1991 the price support scheme was discontinued and wool auctions suspended due to the stockpile of wool that was the result of overproduction.

What this illustrates is that the South Australian sheep producer has problems with indifferent prices and lack of demand for his product, but he has such a low cost of production that he is somewhat insulated from their more devastating effects. He is not inhibited by an uncertain and costly source of livestock feed. Costs associated with the health and shearing of the sheep loom more largely in his mind than the cost of feed.

Returns from livestock grown on hay and grain

The livestock owner in North Africa and the Near East gets much higher prices for his sheep, up to US$300 per well grown lamb in the market, but there is a risk – sometimes the price will drop to around $100. He relies on the milk and meat from the animal for his profit. Wool is not an important source of revenue. He is dependent on outside sources for the major part of his livestock feed and it is this reliance on a high cost and often scarce resource that inhibits him from increasing his flock. Other costs associated with an increased flock are those of shepherding and the construction of a bergerie or concrete shelter for night herding. The difficulty in profiting from sheep comes when the output is related to the cost of production. Boutonnet estimates that from 100 ewes, the Algerian farmer will produce 825 kg of meat per year compared to the 2000 kg of meat per 100 ewes produced in South Australia (Boutonnet, 1989).

In North Africa and the Near East all the costs associated with keeping and feeding the animals is borne by this low level of output. The grazing available on the unproductive fallow and parcours may cost almost nothing to produce, but the nutritional value is practically non-existent. Cereal stubbles on the other hand are reasonable value in early summer because the inefficient harvesting of the crop means that the stubble contains quite respectable quantities of grain that the animals pick up. The major source of feed remains imported feedstuffs and Figure 1.9 shows the cost – $320 000 000 per year for the region. Even when one discounts the feedstuffs going to poultry the costs of production of livestock are very much higher than the cost in Australia where sheep and cattle graze naturally regenerating pasture for almost their total intake and rarely receive supplementary grain.

Table 1.1 *Comparison of the price of 1 kg of sheep meat in Australia, France and Algeria*

Country	Price in $ equivalent	Price in kg. of barley at local prices	Price in minutes at salary of local worker
Australia	1	8	11
France	4	25	50
Algeria	16	70	840

Source: Boutonnet (1989).

An indication of the relative prices in North Africa and the Near East of production for livestock and the effect on the purchasing power of the consumer is shown above in Table 1.1.

Returns from cereals

The prices for cereals received by Australian farmers have generally been poor and at present are as low as they have ever been. World wheat prices are currently 50% in real terms of the price received in 1900 (Tyers & Anderson, 1992). However, the increases in livestock and cereal production shown on South Australian farms in the above graphs reflect the manner in which a low cost system of production provides the farmer with the ability to expand production quickly to take advantage of whatever profit there is. By contrast, when farmers in North Africa and the Near East, who receive prices three times higher for cereals, try to intensify their present system in an attempt to profit from prices that are inflated by subsidies they only continue the pattern of destruction of naturally occurring organic matter that has led to increasing degradation of arable land and their yields continue to stagnate and fall no matter what price they receive for the product.

Exhortations by governments in Morocco, Tunisia and Algeria calling for more wheat as a matter of policy have not been able to overcome this problem (Swearingen, 1988). Equally, the 'Grow more Wheat' campaign in South Australia in the mid-1930s saw yields fall in the district where bare fallow before wheat was the norm and legume pastures were not being used.

Management differences in general

Cereal zone

Of the seven million ha in the cereal zone in South Australia, five and a half million ha are used for cereal and pasture production. As part of the general pattern of the rotation in any one year about two million ha are planted to cereals, three million ha remain in pasture in which medic is predominant, and about 400 000 ha are fallowed – not in a long bare fallow, but a short fallow carried out to clean up a paddock that may have become contaminated with a particular type of weed for various reasons. The tillage and seed bed preparation and the sowing of the seed and application of fertiliser are rapid and cheap. Seed beds have a fine tilth and are level. Weed control is good. Harvesting is efficient and little grain is left behind. Cereal stubble left after harvesting of the crop is sometimes a problem on farms. Sheep eat a lot of it, and the remainder usually rots in the soil during the pasture phase, but if there is so much that it threatens the regeneration of the medic pasture farmers will burn it. Some farmers now make stacks of straw and store it for drought fodder, treating it with the addition of molasses and urea when needed.

Livestock graze the medic pasture (green and dry) and cereal stubble that are produced on the farm. The area planted to cereals and left in pasture is apportioned by individual farmers each year according to market demand and the profit calculated by each farmer to be made from an intensification of either cereals or livestock.

If we take Algeria as characteristic of the region of North Africa and the Near East, we find that about six million ha has the same rainfall pattern as the cereal zone of South Australia. Of this six million ha potentially suitable for cereal production, approximately three million ha is sown to cereals in any given year, and about three million ha is in unproductive bare fallow (Carter, 1974, p. 3).

A year of bare fallow, and the application of nitrogen and phosphate fertilisers are the recommended recipes for cereal production. Farmers sometimes grow intervening fodder or forage crops in place of fallow to provide feed in the form of hay, grain or silage for their livestock (Anon., 1987). Seed beds are undulating and cloddy in spite of the long preparation program. Weed control is poor. Harvesting of grain is inefficient and losses, revealed in an aftergrowth of cereals, are commonly heavy.

Yields on average in Algeria are 624 kg/ha for wheat and 610 kg/ha for barley, while in South Australia from similar soils and with similar rainfall patterns, average yields in the 1970s were 1139 kg/ha for wheat and 1131 kg/ha for barley (Carter, 1974, p. 12).

It is difficult to quantify the numbers of livestock kept on cereal farms in North Africa and the Near East. From observation it appears that most farmers with more than a few hectares keep some sheep – flocks of five to ten are common on smallholdings, and flocks of up to 200 are often seen on more substantial holdings. Some of the larger cooperatives have flocks of 1000 and more, but in all cases the flocks require feedstuffs to be bought in – usually grain, concentrates and hay. The farmer rarely is able to provide more than a small part of the yearly feed requirement from his own farm and the high cost of purchased hay and grain often prevents him from increasing his flock. In Tunisia, Algeria and Morocco farmers interviewed by the authors all gave a normal feeding regime of about 200 g of grain and about one tenth of a bale of hay per sheep per day in winter in addition to grazing the fallow. If cattle are kept it is nearly always on larger farms, in sheds, and on diets of grain and other fodders. A small number of goats are kept in conjunction with sheep.

In Morocco one cow, or sometimes several cattle, are kept on farms of around 10 ha. The animal manure is put out onto the fields before cereals or other crops are sown (Chatterton & Chatterton, 1987–90).

Rangeland

South Australia has about 88 million ha of rangeland, which includes some true desert around Lake Eyre where rainfall is below 125 mm annually. In South Australia only a small proportion of the sheep population for wool production is now kept in this zone. They feed partly on shrubs such as *Atriplex vesicaria* (saltbush) and other *Atriplex spp.* and *Kochia* and bluebush (*Mairena sedifolia*), but mainly on the grazing of annual grasses and other ephemerals and some annual medic that has colonised the region (Squires, 1981). Frequently they are taken to the cereal zone and put on rented pasture when the fragile atriplex and bluebush are threatened by too frequent grazing. It is only high wool prices that keep these flocks economical. Mostly they are not. It is considered by conservationists that these flocks in the rangeland should be removed and the country shut to grazing (Chatterton & Chatterton, 1982, pp. 12–14). This relocation of livestock production to the cereal and high rainfall zones has not been due solely to the eating out of natural pasture in the rangeland, but because of the availability of abundant medic pasture in the cereal zone and sub-clover pasture in the high rainfall country. The greater part of the sheep meat and wool produced in Australia is exported, and oversupply at the Australian end sometimes leads to depressed prices.

In North Africa and the Near East, cereal farmers, driven either by a search for more land (as in South Australia) or by reason of dispossession by colonial powers (in the case of Algeria, for instance), have extended their farming out into the rangeland. These incursions have caused great environmental damage following the introduction to this fragile area of the deep plough and attempts to grow cereal crops year after year (Carter, 1974, p. 4). Where livestock production is concerned, there is a longstanding interrelationship between the cereal zone and the rangeland. Algeria, for example, has 35 million ha of rangeland (Carter, 1975, p. 3). In summer, the flocks (often owned by nomadic tribes) from the rangeland are taken to the cereal belt and the flockowners rent from the farmer the right to graze the cereal stubble – enriched as it is by large quantities of grain lost during the harvesting process. Cereal farmers will sometimes sell the standing crop for grazing when yields are considered too low to harvest. Owners of smaller flocks, some with a little land in the rangeland, will plant small crops of barley or oats for later grazing when there is sufficient moisture available, but rely for their basic supplies of fodder on purchased or home made hay (usually of very poor quality). In spite of sparse feed and only seasonal access to stubble grazing, the nomads have been saved from having to dispose of large numbers of their flock by the importation by governments of grain and other concentrates (usually subsidised), and these now comprise the major part of their livestock fodder. Many programs have been undertaken to revegetate the rangeland but most have not succeeded in producing a substantial amount of grazing for the existing flocks. The demand for meat encourages governments, in response to political pressure arising from high food prices, to subsidise imported grain, but these subsidies do not prevent market prices for animals remaining high because supply is constantly less than the demand. These two factors are preventing any significant reduction of the flocks that graze the little vegetation that remains and thus they are becoming even more dependent upon what is fed 'at the trough'. In Jordan, for example, a study undertaken in 1987 showed that 81% of the fodder required for livestock kept on the range was subsidised and distributed through government agencies (Jamil & Karem, 1987).

The marginal zone

The marginal zone, between the cereal zone and the rangeland, of South Australia is characterised by flat plains originally heavily timbered with mallee (a eucalyptus with multiple trunks and a canopy about two to four metres high), bushy shrubs, grass and some high timber where creeks come down from arid mountain ranges. In some areas there are large expanses of naturally occurring atriplex and bluebush. Where farming has taken place the mallee has been cleared, and when farming has been abandoned there have been invasions of atriplex and bluebush.

The initial incursion into this type of country took place in the latter part of the nineteenth century and although ten years of above average rainfall enabled settlers to farm cereals that produced well, the normal pattern of low winter rainfall and searing summers re-established itself and farmers were forced to abandon their farms and return to the conventional cereal zone. Some farmers went into the area lured by cheap land and easily cleared ground while others were subsidised to settle on farms as a reward for military service. Following the initial intervention of a long cultivated bare fallow on farm land in this zone there was a rapid decline in cereal yields and an increase in erosion. Superphosphate helped briefly but not for long. In the 1950s the medic cultivar Harbinger (*M. littoralis*), which tolerates semi-arid conditions, was made available and this led to some farmland in this zone being clawed back from an environmental disaster. Although external factors such as very high interest rates for bank loans and periods of low world prices for grain and livestock products have in recent times caused these communities severe stress, the farming system based on medic pasture has proved capable of sustaining good cereal yields and pastures for respectable flocks of productive livestock. In prolonged droughts livestock numbers are reduced. A small sum is sometimes paid to the local council to dispose of unmarketable sheep during droughts. A nucleus flock or herd is kept and either hand fed on the farm or sent for agistment (rented pasture) to higher rainfall zones until the drought is ended.

The cereal crop tends to be an irregular one, planted in a good season on land that has been in medic pasture for one or more years. The average yield of such crops is around 1 tonne/ha.

In addition to the farming on the plains, extensive grazing takes place in the wooded mountain ranges of this zone. Graziers rarely have to buy in fodder. They are able (except in prolonged droughts) to maintain their sheep and cattle solely on the productivity of their medic and sub-clover pastures.

The original settlers (in the early part of the nineteenth century) were careful to leave wide shelter belts of mallee trees and bushes around each paddock that they cleared in this zone and to leave areas of natural forest of larger trees as further buffer areas against erosion caused by wind. However, the removal of many of these mallee belts and other natural trees by individual farmers in more recent years, created in part by a government declaration in advance to protect all natural vegetation from further clearing, and in part by greed for more land, has created alarm among those concerned at the threat to this fragile environment. Increasing pressure from environmental groups and concern among neighbouring farmers has led to programs and laws to encourage farmers in these zones to protect trees and to restore to degraded areas a replica of the natural vegetation and forest that originally existed (Chatterton & Chatterton, 1986, pp. 8–11).

In general, marginal land in North Africa and the Near East is characterised by flat, treeless plain and an annual rainfall that varies between 150 and 250 mm. The completely treeless nature of this zone is due to the removal of trees, some say by the Turks in the Near East during World War I, and others say by the search for fire wood over the centuries by the inhabitants. Whatever the reason, one's first experience of this landscape is a sense of shock that land could be so flat and featureless. The degraded mountain ranges that appear here and there are also treeless and stark and offer no opportunity other than for the agriculture of desperation. Re-afforestation programs are now being carried out but for the most part they are only apparent in the form of terraces and small trees.

Livestock production is the major enterprise of the zone, but the extreme grazing pressure of both sedentary flocks belonging to farmers with individual landholdings and the nomadic flocks that periodically cross the land – notably in the summer *en route* to the cereal stubble in the cereal zone and again at the beginning of winter when they return to the desert – has reduced the vegetation to a few annuals and shrubs unpalatable to animals, and most of the ground is bare. Nomadic flocks consist mainly of sheep, but there are in addition large numbers of goats and camels that make the passage. If the rangeland pasture has not been particularly productive in winter (and it rarely is these days) the flocks are brought into the marginal zone in the spring and rapidly consume whatever vegetation exists outside the farm boundary. They then camp in the zone and are fed hay and grain until it is convenient to move on. Sedentary flocks are dependent on whatever grain (usually barley) can be grown for grazing, poor quality straw made by the farmer, and hay and grain, the latter mostly available at subsidised prices from government sources. These various fodders are for sale in the sheep market that operates from the nearest large centre on a regular basis. Not all the marginal zone is subject to nomadic transhumance, as this seasonal migration is called, and the transhumance that exists is carried out over pathways that are traditional and defined.

Cereal production in this zone has begun mainly for social reasons over the past fifty or so years. Cereal crops were grown as farmers tried to gain rights over common land, or having been forced from their farms in better zones were trying to re-establish themselves as farmers. In more recent times (for instance Algeria in the 1970s) there have been unwise government schemes to settle nomads and other persons in need of political pacification on this marginal land. In all cases, the land has been quickly exhausted by continuous cropping with cereals, mostly barley, during which deep ploughing takes place to depths of 30 cm and frequent passages are needed to prepare seed beds. Farmers in this zone rarely use nitrogen or any fertiliser because their returns are too low to enable them to pay for it. They cultivate neither pasture or forage crops. Landowners with less than 10 ha cultivate their land with mule and wooden plough (there are large numbers of farmers like this in Morocco), and some with intermediate sized holdings (between 10 and 30 ha) employ the government machinery centre or private contractor to prepare the soil for sowing using tractors, deep ploughs and other implements used in the long tillage program. In parts of Jordan and Algeria there is a growing tendency to cultivate the soil with a tandem disc, then a harrow, and sow directly into this seed bed – sometimes with phosphate fertiliser. The sowing of seed and the application of fertiliser are by hand. While this costs less and saves time, it does not increase production or prevent erosion. Cereals sown on the degraded plains year after year rarely yield more than 400 kg/ha in the best years and usually only about 150 kg/ha. The crop is frequently grazed by animals as a standing crop. At other times it is reaped by hand for grain for family use. There is a noticeable contrast between the South Australian marginal zone, with increasing pasture and cereal production and erosion under control, and that of the Near East and North Africa where yields of cereals and pasture are declining and erosion has reached disaster levels with no sign of a reversal.

Soil fertility

The fundamental base for good yields remains the degree of fertility in the soil, whether natural or induced. Gintzburger studied that most forbidding of all dry farmland, the rangeland. He applied superphosphate to the soil and found that the subsequent encouragement of the naturally occurring legumes (in particular the medics) markedly increased the productivity of this unpromising territory (Gintzburger, 1980). If superphosphate and legumes can have such a dramatic effect on degraded rangeland is it any wonder that they can enhance the farm productivity in the cereal zone where the existing soil structure and better rainfall provide more encouraging circumstances? South Australian farmers who found medics and sub-clovers on their land used them to evolve a farming system within which a cycle of fertility is established and continues to underpin productivity. The management of the medic system gives it an edge over other legume crops when it comes to sustainability of soil fertility. The ability to fix nitrogen in the soil is common to all legumes but in the case of lentils and chick peas much of the nitrogen produced by the legume is removed when the crop is harvested. In the case of vetch and other forages that are cut for hay, most of the nitrogen is removed with the hay. When animals graze vetch the organic matter and nitrogen are returned to the soil as is the case with medic, but vetch and forage peas must be resown each year and the cost of this and the potential environmental damage due to extra cultivation is a disadvantage. Dryland farmers in Morocco grow maize over the summer but costs involved tend to outweigh any benefits. Some small cereal farmers in North Africa and the Near East have adopted a system that includes growing forage legumes for their dairy cows, but they are not always prepared to do so for sheep.

The existence of indigenous medics in North Africa and the Near East

Obviously if a medic farming system is to be used in the region, there will need to be present medics suited to the ecological conditions on most farms. In one survey alone – along the Libyan littoral zone, Gintzburger and Blesing (Gintzburger & Blesing, 1979) identified over a thousand ecotypes of medic, and in the Northern Syrian rangeland Cocks (Cocks, 1984, pp. 288–91) identified another thousand. Adem (Adem, 1989, pp. 227–8) in Algeria found many ecotypes of naturally occurring medic above 600 m and higher and out into the rangeland. It is not difficult to find medics growing in most of the region. When we first began prospecting for medics in the region we were impressed with the prolific medics that we found growing on the sites of ancient Roman cities where they had not been threatened by cultivation or overgrazing. Subsequently we found them on village common land, growing between plots on research centres, and on roadsides and uncultivated parcours (rough grazing land). It is notable that they have almost disappeared from cultivated farm land because of deep ploughing and the long bare fallow. Another factor has mitigated against the medic. This is the use of nitrogen fertiliser on cereal crops, as it has encouraged grasses and suppressed the leguminous plants. The absence of abundant medic in the rangeland is due both to the fierce overgrazing and to the incursion of cereal growing in the marginal zone, although formerly medic provided a large part of the natural pasture. In a few areas the assault on medic has been so great that the bacteria necessary for the fixation of nitrogen by the plant has disappeared and must be artificially re-introduced if medic is to be re-established successfully (WAOPA, 1985, pp. 23–7). The question of producing supplies of seed for both farmland and rangeland programs is being tackled and this will be examined further on. The ubiquitous nature of medic in the region may well mean that a change in cultivation methods and the application of phosphate fertiliser will be the major method of restoring medic to the farming system.

Programs to reintroduce medics to farm and rangeland

In order to enable farmers to change their present system to a medic farming system on dryland farms and rangeland in the region three major points will need attention. They are as follows.

- The profitability and productivity of the system needs initially to be made known to farmers and livestock owners by means of demonstration of two critical components of the system (see below) together with a comparison of the costs and returns of the existing and the proposed system, if possible, on their own farms.
- Management of the medic pasture in order to exploit it adequately and to ensure its regeneration requires that the number of sheep grazed must be related to the pasture available. This ratio varies according to the rainfall zone. For example, under moderate rainfall conditions, one can assume that a farmer owning five sheep will need one hectare of pasture to feed them for a large part of the year.
- To safeguard the medic seed bank for regeneration after the cereal phase, farmers must change from deep to shallow cultivation when preparing the seed bed during the cereal phase. This is a fundamental change requiring education, demonstration of the savings in cost and energy and the provision of properly designed implements.

The analysis of attempts to introduce the system to North Africa and the Near East that follows will undertake an examination of the political economy of dryland farming as well as the technical and institutional instruments associated with the discovery and establishment of a medic farming system in Australia and, later, with the introduction and partial adoption of it in the Near East and North Africa.

2
Farming in South Australia before medic

Introduction

The farmers who came to South Australia from England in 1836 and later were forced to discover new ways of tilling the soil, harvesting cereals and keeping livestock.

Robert Chambers has argued that the experience together with the common sense and innate wisdom of farmers should be acknowledged and used by those who plan and operate farming projects in the developing world. Chambers has written scornfully of the 'development tourist' – the technical expert armed with a diploma or degree in agricultural science who goes from development projects in one climatic zone to another, emerging from each as an 'expert' on yet another aspect of Third World agriculture – and claims that this is a major reason why many projects destroy rather than improve the existing agricultural resource (Chambers, 1983).

The other side of that argument is put by Dr. P.E. Cocks (formerly responsible for the pasture research program at ICARDA in Syria and consultant to a number of Australian dryland farming projects in North Africa and the Near East). *A propos* of the suggestion that Australian farmers were better at transferring the knowledge of their medic farming system to Arab farmers than technical experts he said that 'Australian farmers are no better at understanding (the farming system) than Australian scientists. . . . There is no brotherhood of farmers and the sooner we get rid of this idea the better' (Springborg, 1985, p. 26).

These two opinions about the farmer's role in agriculture are pertinent to the conflict that is charted in this book between the farmer and the technical expert. The challenges of dryland farming in Australia soon focused on the need to find some form of rotation that would provide a stable underpinning to cereal and livestock production. The farmers' search for a solution began and ended on the farm and grew out of bright ideas and practical innovations developed by individual farmers and their neighbours. The search of the technical expert took place on the research centre where attempts were made to adopt ideas that came from international sources.

Land use prior to European settlement

Prior to the coming of English settlers to South Australia in 1836 there was no farming. The Aborigines who were, until the invasion of these Anglo-Celts, the only occupiers of the vast territory and who lived there undisturbed for thousands of years, did not cultivate the soil nor did they keep livestock. The women were hunter-gatherers of roots, seeds and fruits and the men hunted for animals and fish. While tribes had recognised territories, they were nomadic within those territories and did not build stone edifices or monuments. Their shelter was light bush cover and their monuments were the natural rocks and lakes around them. Their cave paintings represented a history of nomadism, hunting and gathering, and a mythologised dreamtime which bound them, their ancestors, the birds, the animals, the fish, the land and the sky together. Their culture was regarded as primitive by the invaders and the sophisticated complexity of their relationship with each other and the environment remained unknown or unacknowledged by successive white Australian Governments until late into the twentieth century.

Once the land had been glimpsed by the British explorers, Flinders from the sea and Sturt from the River Murray, on exploration journeys and 'acquired' for Britain, the right of the Aborigine to continue the relationship with the land was swept aside. In the minds of the discoverers their 'ownership' of this new land only needed to be enforced by the arrival of British settlers under the aegis of the British Government or its agent to become *de jure* as well as *de facto.* These settlers would, in turn, establish their individual *de jure* right to own the land by buying it off the British Government.

The origins of the South Australian settlement scheme

A group of young idealists, led by Edward Gibbon Wakefield, had persuaded the British Government to become patron of a settlement scheme that would reflect enlightened attitudes of social equality, civil liberties and religious freedom. Their objective was to create a colony based on orderly settlement. As far as was possible each settler would buy for cash exactly the same number of acres, settlement would begin from the centre and move out as communities were created, not simply as demand for land grew, or idiosyncratically. Each new incursion into the wilderness would be first surveyed and then sold off in the approved unit of land. This orderly settlement was to ensure that communities were properly served with roads, schools and other services proper to a self-respecting community. The settlers under this scheme would not be opportunistic riff-raff and entrepreneurs who employed slave labour, speculated in expropriated land and exploited their more modest and ingenuous neighbours. Such behaviour was commonly believed to characterise many of the colonial settlement schemes of the previous century. Wakefield wanted his settlers to come from that respectable yeoman stratum of British society that would pay for their new land, employ honest British workingmen and women and produce a colony with respect for God and associated civilised virtues (Pike, 1957, pp. 77–81).

The British Government agreed to accord the scheme recognition and some degree of protection provided it did not become a charge on the British Treasury. Land sales were to underpin the Colony's economy until the export of agricultural produce could take over. A Commission was appointed to administer its operation and a Governor appointed to oversee the interests of the British Government. Blocks of land surveyed for farmland and township lots within the proposed city of Adelaide by the Commission would be sold to the settlers. This would bring in cash which would then be used to recruit more men and women from England who would sail for South Australia to work as farm labourers, apprentice tradesmen and domestic servants until they had saved enough capital to enable them in their turn to become 'yeoman farmers'. Unhappily, the first phase of this financial scheme came to grief. The surveys were delayed for two years due to inefficiencies in the organisation of the settlement's food and other requirements and squabbles about the site of the city of Adelaide. The Commission was required by the British Government to have sufficient cash reserves within two years of the commencement of the scheme to justify the protective role of British law. A rival group of roving British capitalists who set up the South Australian Company played on the fears of the Commission that it would not sell land quickly enough to keep the cash flowing. As the two year limit loomed and sales were far from sufficient because of the delays in the survey, the South Australian Company was able to buy in a 'job lot' at rock bottom prices a large proportion of the alienated land to be surveyed, and also the further promise of large, cheap leases of the heavily timbered, hill country grazing land that surrounded the original coastal plain on which the settlement began. In addition to its inefficiencies in carrying out the initial organisation of the settlement and the surveying, the Commission had when planning its finances given no thought to a fund to cover administrative costs. This led to the bankruptcy of the Colony in 1841 that caused financial hardship to many of the early settlers who had contracted services to the Commission (Bull, 1878, pp. 139–40). The British Government, faced with either abandoning the colony altogether or letting it slide into the hands of the South Australian Company and its principals, stepped in and took over and made some loans available for consolidated revenue. The South Australian Government was born in 1843 when an advisory Legislative Council was appointed (Bull, 1878, pp. 212–3).

In the meantime, some of the original settlers had finally been able to take possession of surveyed farms of 32 ha each, plus rights to common land for grazing and had begun to farm (Williams, 1974, pp. 67–71).

The Aborigines were consigned to the wilderness, dispossessed and hunted. When they fought back they were subjected to humiliation, imprisonment and death. Later they suffered from misguided missionary zeal and were the subject of disastrous government policies that shattered families and most of what remained of their human dignity.

The first farmers

As the Wakefield scheme got under way, little attempt appeared to be made to recruit experienced farmers from England. Out of the first 4454 settlers to arrive, only 57 gave as their occupation as farmers, 455 were registered as agricultural labourers, 83 were shepherds and two were agricultural machinery makers (Capper, 1838, pp. 196–8).

Colonel William Light (the Commission's initial surveyor for the settlement) wrote in his diary in 1836 that agriculture was obviously to be the basic economic resource for the colony and that while no looms or engines were sent out 'many immigrants brought ploughs, harrows, etc. (and) it appears therefore that agriculture was most thought of by almost every immigrant – and I think by the regulations drawn up ... by the Commissioners themselves' (Light, 1984, p. 100).

The inducement offered to those who showed an interest in emigrating included the bravado that for success in farming one needed only 'a taste and fondness for country life' (Capper, 1838, p. 130).

In spite of Light's belief that farmers brought their implements with them, the exhaustive list of necessities for emigrants issued by the Commission was concerned almost exclusively with household goods and equipment and the only tools recommended for passage on the ship were those suitable for rudimentary gardening and household work and repairs. British agricultural machinery manufacturers were quick to flood the Colony with advertisements for British made implements for sale and shipment to the settlers. The equipment listed was that used on British farms (Capper, 1838, p. 3).

English farmers used a crop cycle in which cereals were sown in the spring as the soil began to warm up. Cereals grew throughout the warm, almost constantly moist summer and were harvested in early autumn. Rotations consisted of perennial grasses and clover, turnips, mangolds and kale – all relying on high levels of moisture and, in the case of the rootcrops, a great deal of labour. Hay was made for animals in winter because it was too cold then for pasture to grow and snow, frost and sleet throughout winter made it necessary to shed animals for long periods – particularly during the lambing and calving season. The steel plough, used to prepare the soil for cereal and other crops, was being perfected to go deep and to open and then invert the soil so that weeds were buried to kill them and moisture was able to evaporate from the sodden soil. During the early nineteenth century there was an excess of agricultural labour available in Britain (it was this grave unemployment that had been one of the reasons to welcome the Wakefield scheme, as it provided somewhere to send these superfluous persons) and crops that required hand hoeing, harvesting and storing posed no problems to the farmer.

Farmers who came from this background were shocked at what they found in dry South Australia.

The climate and its effect on British farmers

From the moment the farmer settlers arrived in 1836 there was pressure on them to produce food. The administrative bungles and the frightful climate that plagued the settlement made this difficult. Surveyor Light bitterly records that during his first three years in the colony of South Australia he had to walk everywhere from the day he arrived, he had no means of transporting his kit or gear except by hand cart or wheelbarrow and saw his staff suffer from scurvy because of a seven month long diet of saltpork and no fresh meat or vegetables. He had seen to it that a garden was planted at Rapid Bay where he had first settled his men on their arrival in August 1836, but four months of miserable weather – gales every week, much rain, both searing heat and miserable cold being experienced on the same day – was the result of southerly winds playing havoc with the arid summer climate. This was not conducive to productive gardening and when the rapid increase in daily temperature that accompanies the hot dry summer in South Australia hit the gardens little was left to salvage. By November, Light was forced to send a ship off to the nearby Van Diemen's land for cattle for stock and provisions, and four months later in February 1837 the Commissioner sent another ship to Sydney in the Colony of New South Wales 'to buy cattle for food, working bullocks and wagons'. These colonies were enjoying the advantages of virgin land and, in the case of the mainland states, large tracts of natural pasture. However, they were concentrating on supplying their own settlers and gaining export markets and therefore asked high prices from their neighbouring colony for whatever was surplus (Bull, 1878, pp. 19, 49, 55–7, 135, 170). Light wrote that at times his survey staff spent days with 'hardly anything but biscuit, sometimes not that' and had to stand by while their small survey truck was requisitioned to carry the baggage of settlers from the port of Adelaide up to the new settlement. The food available to the survey team came from the Commission's stores and as the new settlers were without any resources of their own in the way of food, these stores had to be spread very thinly in order to share out what was available (Light, 1984, p. 115).

Opinions about the climate and its effect varied considerably. One farmer wrote that 'no one could make a little England out of such a land' (Meinig, 1963, p. 16) and another early settler in 1841 passed to his family the 'generally received opinion that the climate and soil of South Australia was uncongenial to the growth of wheat, barley, potatoes, etc., due to the failure to raise crops during summer' (Price, 1924, p. 116). An English settler named Milburn also wrote home saying 'the climate ... though so fine is hot and incalculable suffering often comes from want of rain. To an European, the scenes which sometimes occur are beyond conception ...'. He was not dismayed, however, and ventured to suggest that 'the climate, hot as it is and dry in some seasons seems to be suited to growing fleeces' and guessed that South Australia should one day be able to carry about 12 million sheep and draw in an income from wool of about $AUS 5 200 000 (Capper, 1838, pp. 33–6). Capper, an early historian, praised the climate and another writer, Forster, called it bracing and healthy and gave it credit for the low number of chest and other complaints among the settlers (Capper, 1838, p. 43; Forster, 1866, pp. 37–8).

In 1904 the Commercial Agent for New South Wales told the Royal Colonial Institute that 'There is this to be said in favour of drought – lessons may be learnt – it rests the ground' and in his opinion another thing in favour of Australia was that as 'there is no winter as it is understood in

Britain and North America . . . (while) . . . it means nothing for the growing of wheat . . . it is good for the production of cattle and butter because there can be open grazing all year round' (RCI, 1904, p. 80). Another delegate, Mr. F. Storie Dixon, gave his opinion that in Australia 'climatic conditions from time immemorial have been most repulsive' (RCI, 1904, p. 103).

This 'repulsive' climate consisted of long, hot, often totally rainless summers, relieved by a winter rainfall that was difficult to predict. Temperatures in winter could be as low as 1° C at night and rise to 15° C during the day. In late spring, summer and often early autumn, nights could be as hot as 28° C, with temperatures increasing to 42° C during the day. At other times in spring and autumn, the range could be as low as 1° C at night and as high as 37° C during the day. At the same time, temperatures could fall 20° C in a single day during summer and rise by as much again within 24 hours. Winds from the desert in the North were searingly hot and those from the Southern Ocean, bitterly cold (Capper, 1838, Forster, 1866).

Settlers found that even the more fertile areas available to them (and these were very few – mainly in the Adelaide Hills to begin with and later in the Lower South East) rarely received much above 500 mm of rainfall per annum and most of the initially arable land received between 350 and 450 mm per annum. Droughts even in the better regions were common and even now it is a rule of thumb that drought of greater or lesser intensity will occur every five years.

The only reliable river in the State was the River Murray with its headwaters in Victoria and New South Wales. It was not an easy river to divert for irrigation, subject as it was to frequent flooding and unpredictable drying up when the largely unexplored headwaters a thousand miles away were afflicted with drought. It became a transport channel for river boats that took export wool and grain from the North Eastern farms to the ships berthed at Goolwa and eventually was transformed by a series of locks and weirs and a reticulated system that made it the State's major source of water for industrial and domestic use.

Charles Fenner, an early geographer wrote that 'the settlers came to a land that was not only unknown to them so far as relief and climate were concerned, but it was practically unknown to the world' (Fenner, 1931, p. 13).

The scarcity of labour

An immediate problem the farmers faced was the lack of labour. Under the conditions of the Wakefield scheme they were supposed not to employ aboriginals as farm hands for fear of the scheme being tarred with the 'slave labour' brush that Britain was newly eager to escape. The demand for skilled workers to build houses, erect post fences, perform civil engineering works, and provide personnel in the commercial and service sectors was great, but the reservoir of such workers was so small that wages were very high (Bull, 1878, pp. 200–4). In spite of the interdict against employing aboriginals, some of the graziers on the leased outstations did employ them as shepherds and rouseabouts, but most farmers, more under the eye of the administration, either paid high prices to the British workmen available or did without until convicts (most of whom were escapees) from the Eastern states became plentiful (Bull, 1878, p. 181). German settlers under the leadership of Kavel (paying the capitalist landholder George Fife Angas dear for the opportunity of buying sections from his special survey in the Mt Lofty ranges) came out in 1838 and tied their women to the plough and even used them as shearers – tying the toe of the woman to the sheep's leg while the man clipped the sheep of its wool (Bull, 1878, pp. 106–7). Thankfully, the German settlers were eventually able to afford horses. Even when convicts (freed and escaped) became available they were not particularly trustworthy or able employees, and farmers did not like employing them. In the 1840s when money was scarce it is claimed they took to stealing in lieu of wages (Bull, 1878, pp. 99–101). The agricultural sector has never been a large employer of labour in South Australia in spite of the relatively large sizes of modern farms of 600 ha and more now used for cereal and sheep production. The tendency that evolved from those early days to rely on the family labour for the farm has continued until the present day. Few farmers have a permanent workman, or have a tradition of having had one. Sheep are left in fenced paddocks to graze untended, and mechanical harvesting of cereals together with the rapid tillage possible with shallow cultivation and combine seeders mean that farmers only employ professional workers during the sheepshearing season.

Shallow cultivation

The practical difficulty the South Australian farmer settlers faced in providing adequate supplies of food (meat, milk, grain, vegetables and fruit) for their own families and those of the other settlers rapidly pouring into the new colony becomes apparent when one considers the physical effort needed to clear the scrub, break the soil, plant, weed and harvest without any machinery and precious little labour in a climate that is harsh and unforgiving. The farmer had not only to think of his family and the settlement's needs. The future of the colony depended on farmers and graziers being able to produce sufficient wool and wheat for export markets to provide a continuing source of finance for the settlement.

Some idea of the physical energy and accommodation to the climate required was given by H.E. Dodd who, in 1790, was provided with 22 ha in New South Wales. Dodd's first attempts to grow cereals failed because he tried to plant and harvest according to the British calendar. He began again, this time sowing the cereals not in spring, as in England, but in mid-winter so that he could harvest the crop in early summer before the hot dry summer winds burnt it up. His land had to be sown by hand. He used convict labour to prepare the seed bed. He required each man to hoe 400 sq. m each day, and as this was a great amount, in some places the earth, which was very hard, was just scratched over. He sowed seed at the rate of 140 kg/ha and noted that seed planted about 8 cm deep germinated best. After the seed bed preparation he left the land 'open for some months to receive benefit from sun and air' – and one can assume from this that he had had the hoeing done in late summer or early autumn and was, in effect, putting the land into a type of long fallow. In the intervening period bushes and trees were burnt and the ash dug into the open soil in the belief that this would sweeten the soil. Dodd was very anxious to find some manure for the soil and says how he 'yearned' for some animal manure, but none was available. He was tolerably pleased with the subsequent wheat crop which he expected to yield just over 900 kg/ha but the oats was 'in ear though no more than six inches high' and he did not expect it to return its seed. The barley was 'little better than the oats' (Callaghan & Millington, 1956, pp. 12–20).

The South Australian farmers bought farms of about 32 ha that had to be fenced and on which a house and farm buildings had to be built, and all this in addition to sowing crops and looking after livestock.

The solution to their critical lack of labour and draught animals appeared to lie in finding some mechanical means to clear and work their land. In the process they found they did not need to plough the soil 30 cm deep in order to prepare a seed bed. The result was the same if the soil was only scratched to 10 cm and the seed lightly covered and left to germinate. Shallow cultivation was an invention of necessity, but it proved to be a most remarkable and valuable farming innovation.

Clearing the land

The wide, open plains surrounding Adelaide were covered with tall trees but not a great deal of bush. Most of the initial clearing of the land for houses and crops was done by teams of men using axes and hand saws. When the settlers moved out into the plains to the north of Adelaide and even further onto Yorke Peninsula and Eyre Peninsula they found clearing more difficult. Tough mallee trees covered the ground thickly. Even today one can see on the Blanchetown Plains of the Murray Mallee and in parts of the Upper South East (the last regions to be settled for farming) thousands of hectares of heavily wooded country – mallee trees with their continuous canopy of dull green and bronze leaves, the multiple branches of each tree creating a thick screen preventing one seeing any distance ahead, and the thick knobbly roots clinging to the soil with tenacity. Sheep graze beneath the canopy, eating the ephemerals that appear after the sparse rain as well as the saltbush (*atriplex* spp.) and bluebush (*maireana* spp.) that have colonised where the mallee had been initially cleared for wheat crops and later abandoned. A million or more hectares of this mallee was cleared during the expansion of the wheatlands in between 1860 and 1880. Settlers and townsfolk benefited from the warmth of the mallee roots, which proved excellent firewood for cooking and heating, and when finally mallee roots became scarce (not until the 1970s) what remained to be salvaged became expensive and much sought after (Williams, 1974, pp. 166–90.

The need to clear this land cheaply and then to get a crop in quickly, led to two farmer inventions. One was for the clearing operation, and the other was for the mechanisation of the process of shallow cultivation in order to cultivate the soil thus cleared.

In 1868 a farmer called Mullens at Gawler developed a heavy roller drawn by horses that pushed over the mallee trees and left behind exposed logs and trash that was later burnt. The ground was opened up with a spiked log and cereal seed was then broadcast. When the crop had been harvested the brush and any new shoots were burnt again with the stubble and a further crop sown. Some farmers thought this burning 'sweetened' the soil and they claimed it did instead of manure. Many farmers cleared their land with the Mulleniser as it was called, but the problem still remained of the knobbly roots that had to be dug out.

Between 1876 and 1878 several farmers, (among them Stott, Scobie and Spry) invented various types of grubbers, but most farmers were still forced to use various hand held tools to get the stumps out (Williams, 1974, pp. 143–50).

The price of firewood helped provide an incentive to grub out the stumps – particularly during droughts and other rural reverses. Settlers who bought land in these marginal zones were required, as part of the conditions under which they were granted their perpetual lease of the land, to clear a percentage of the land each year.

In 1876 R.B. Smith, a farmer of Kalkabury, decided not to pull out the tree stumps remaining in the ground after clearing but to evade them. He took the mouldboard plough and hinged the beam shares so that they rose gently out of the ground when coming up against a stump or root. Weights placed on the extensions of the beams pressured the plough share back into the ground as soon as the stump or root was passed. In time, the passage of the plough got rid of all except the largest roots and stumps. This was the celebrated 'Stump Jump' plough that, together with the Mulleniser, enabled farmers to move out into scrub and heavily timbered country, clear it, and make it pay because it could be cleared cheaply and sown cheaply. The cost of grubbing had been from $AUS 15 to 20 per acre, but a three furrow stump jump plough that did away with the need to grub could be bought for $AUS 46 (Williams, 1974, p. 149).

In 1881 a total of 430 000 ha had been cleared for farming. In 1885 the scrub leases that had been granted at concessional rates by the government in recognition of the difficulties of clearing and farming scrub land were discontinued and scrub land was for purposes of sale and regulation treated like any other (Williams, 1974, p. 151).

Today the mallee forests are no longer cleared with impunity and since 1979 the government has been trying, with varying degrees of success, to perfect legislation that will encourage farmers to leave existing mallee as conservation reserves (Chatterton & Chatterton, 1986, pp. 9–11).

The implements used for shallow cultivation

The plough

Once the land was cleared the operations necessary to prepare a seed bed occupied the minds of the farmers. They quickly found that the ploughs brought from England were not much use in South Australia because the soil was shallow and very stony. The English plough was designed to cut into the soil with the ploughshare and coulter, and then to lift the soil and invert it with the mouldboard. This opened the wet soil to the air and also covered weeds and buried them so that they died. In South Australia, the problem was to conserve moisture, not to dehydrate it, there were few weeds due to the relative infertility and shallowness of the soil, and the obstacles created by stones and tree stumps were better avoided by a shallow plough. As well as making the plough smaller and the mouldboard much wider so that it spread the soil and did not invert it, the South Australian farmer fitted wheels to prevent it entering the soil to a greater depth than 10–12 cm. Weeds, uprooted by the tillage, were left on the surface of the tilled soil and in the dry air they withered and died. As horses and some oxen became more plentiful in 1850 due to the demand for traction animals from the newly developing mining industry, farmers joined several ploughs together on a frame and yoked teams of horses together. This became the first widespread form of mechanised shallow cultivation. In one of the first signs of the laconic but not lethargic stereotype of an Australian farmer, a seat was mounted on the plough frame and Australian farmers sat royally behind their horses, instead of following the plough on foot as did their English cousins. The first tyned scarifier – that eventually replaced the plough as the common tillage implement – was demonstrated by a farmer in 1853.

Nonetheless, many of the early South Australian farmers were still using spades to cultivate the soil on their 32 ha as late as 1850 according to replies to a questionnaire. They claimed that ploughing was 'a major difficulty'. The cost of contract ploughing with bullocks was \$AUS 9.90/ha prior to 1850, and although the price fell to \$AUS 4.47/ha after this, it was still a high price to be paid when wheat was \$AUS 2.45/100 kg and yields were around 1120 kg/ha (Pike, 1957, p. 330).

When the scarifier became available and commonly used later in the century, one farmer could cultivate 4 ha per day (Callaghan & Millington, 1956, pp. 315–26).

The Australian scarifier

In 1881 a stump jump scarifier was invented by a farmer called Stott and widely adopted, and this, together with a mobile rake for piling up roots and branches on newly cleared land, gave land settlement a huge boost. The scarifier was a radical departure from the heavy steel plough. It consists of hinged curved tynes bolted to a heavy steel frame with springs to hold them in position. The tynes are about 3 cm wide and about half a metre long. Each tyne is fitted with a point which is bolted directly to the tyne. The points can be narrow (10 cm) or broad (15 cm). The narrow point (Figure 2.1, left) is used to open soil when conditions are particularly dry and the soil is particularly hard. The broad point (Figure 2.1, right) is then used for the secondary cultivation. The points open the ground but do not create the large clods left by the European deep plough. On large farms bigger scarifiers are trailed and bigger tractors (around 100 h.p.) are used.

Figure 2.1.   Photographs of points. The narrow point (left) is used for breaking open the soil in the first passage in autumn. The broad point (right) is used for weed control in subsequent cultivations. (Photos: F. Botts.)

In spite of concerns to the contrary, using the scarifier does not create a hard pan over time. A scarifier pulled with an adequate tractor can open previously uncultivated hard stony ground, and can even break up bitumen covered ground without difficulty. A small thirteen tyne scarifier of three rows on linkage can be pulled behind a small wheeled tractor of about 45 h.p. The spacing of the tynes is important to the efficiency of the scarifier as an implement for tillage and weed control. The tynes of the scarifier are 17 cm apart and a depth wheel is fitted to ensure that the implement does not till deeper than 10 cm. The spacing of the tynes ensures that all the soil covered by the implement is tilled. Weeds are pulled out of the ground and left on the surface to dry and die. The scarifier is of simple construction, has few moving parts, and a supply of replacement points is easy to keep on hand and require only bolting off and on.

It is cheap to buy and the ease and economy of its maintenance, and its ability to rapidly prepare a seed bed, endeared it to South Australian farmers and they eventually discarded the Australian mouldboard plough in its favour.

The Australian scarifier is often mistakenly confused with the chisel plough used in Europe and North America (Figure 2.2). This chisel plough (known sometimes as the 'Canadienne') has been designed to plough deep (25–35 cm). It usually has narrow points – about 5 cm – and it has a tyne spacing of about 30 cm. If used to cultivate at a shallow depth, its design prevents it from cultivating all the ground and so weed control is not effective.

The Australian scarifier, to European eyes, is also similar to a light cultivator (known as a 'ducksfoot cultivator' in Syria, for instance) as far as its tyne lay out is concerned. The Australian scarifier is much more robust and powerful and its heavy spring pressure enables it to enter the most difficult ground with ease.

Figure 2.2. Photographs of (top) scarifier, showing heavy duty springs and tyne spacing; and (bottom) chisel plough, showing wide tyne spacing and poor cultivation of soil at shallow depth. (Photos: F. Botts.)

The Australian disc plough

A disc plough was also invented for shallow cultivation. Here again there was a striking difference between the disc plough of Europe and North America and that of Australia (Figure 2.3).

The European disc plough has large discs offset on a frame that ploughs to depths of more than 25 cm and in most cases needs a crawler tractor to pull it. A common configuration is a three metre disc plough with nine discs. Some smaller European disc ploughs are linkage mounted.

The Australian disc plough was fitted with a much larger number of small discs offset on a frame and was trailed. A common configuration is a three metre disc plough with fourteen discs. Australian disc ploughs are not linkage mounted as the scarifier often is. The Australian disc can be pulled behind a 65–75 h.p. wheeled tractor.

The disc plough for shallow cultivation is not widely used on South Australian farms. In Western Australia and New South Wales, where farmers still tend to grow wheat on wheat, the disc plough is used on some properties to cut up the stubble, but in South Australia during the medic phase the stubble rots down and is returned to the soil as organic matter when the ground is cultivated with the scarifier during the subsequent cereal phase. The cost of purchase and maintenance of the disc plough is high. The bearings, which are very robust, are expensive to replace and in isolated areas are often not readily obtainable at the critical moment. Initially some farmers bought disc ploughs for hard ground and excessive stubble, but improvements in the robust nature of the scarifier have removed this advantage.

The combine seeder

The cost and time involved in preparing a seed bed, sowing seed and applying fertiliser continued to focus farmer attention on ways and means of providing a more efficient mechanical operation. Farmers and manufacturers combined the scarifier with the seed drill. Boxes were mounted above the scarifier for cereal seed, superphosphate and later small legume pasture seed, from which extended tubes enabled seed and fertiliser to be placed precisely in the soil together. Harrows were attached to the scarifier and dragged behind so that as the seed and fertiliser fell into the soil both were lightly covered by the harrow. With this combine seeder, as it became widely known, the farmer can prepare a seed bed, plant the seed, place the fertiliser, and cover the seed – all in one passage over the ground (Figure 2.4). Thus Australian farmers perfected what must surely be one of the

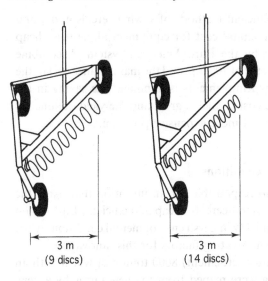

3 m
(9 discs)

3 m
(14 discs)

Figure 2.3.   Two types of disc plough, showing different spacing, for deep tillage (left) and shallow tillage (right). (Source: Thomas, 1980.)

Figure 2.4.   Photograph of combine seeder, showing seed and fertiliser boxes, tynes in operation and covering harrows. (Source: South Australian Department of Agriculture.)

cheapest, most rapid, and efficient methods of sowing cereals in dryland conditions in the world. In capital cost for equipment alone it is cheap compared to that required by the Euro/American system. A combine seeder can be drawn by a tractor of 100 h.p. and the maintenance cost of the robust and simply engineered implements is minimal. The saving in the farmer's time as well as the cost of fuel is great and the yield obtained is equal to that of the longer and more expensive program.

Harvesting cereals in dry conditions

Farmer innovation was also responsible for an invention that made the mechanical harvesting of dryland cereals cheap and efficient. Little or no labour and the need to act quickly in harsh environmental conditions when the crop was ripe were the motivating factors for this innovation.

In 1843–4 (seven years after settlement) 8000 tonnes of wheat with an average yield of 2100 kg/ha were reaped from the new farms by a new, horse drawn machine that passed through the crop threshing the grain and chaff from the heads of the wheat while leaving the stalks uncut in the ground. The machine was invented by J.W. Bull, who was one of the first farmers in the state. He had a good crop of wheat in 1842 but because he found it impossible to get labour to harvest it when it became ripe in early December he faced heavy losses. It was not until 24th December that he was able to persuade some workers, who had until then been receiving high wages in the town, to come to his farm to take the crop off. The acute labour shortage had required soldiers to 'lay down their arms', and 'soft handed gentlemen' had been forced to join them to harvest the bumper crop on many farms (Bull, 1878, p. 237). Bull took his five labourers, and agreed to pay them $AUS 1.50 in cash and a bottle of rum per acre together with their rations. He allowed them the day off on Christmas Day and supplied them with Christmas dinner. A hot dry wind began blowing on Christmas Day and on the following day, Bull expected that his workers would get to it. But they did not. He found them in the grog shop. By the time he sobered them up and persuaded them to start harvesting, the wheat had lost over 70 kg/ha and because of the over-ripeness of the ears, more went in the harvesting. As prices for wheat were low and buyers scarce, Bull was despondent. While demonstrating to the workers the way in which he wanted them to carefully reap the wheat, he was struck by an idea for mechanical harvesting and threshing, and spent the next few months drawing out his ideas and constructing models. The principle was that beaters would thrash the grain out of the ears leaving the plant standing.

Most of his friends ridiculed his idea, but a group of cronies, who had formed the 'Corn Exchange Committee' to help farmers who were suffering the difficulties of harvesting, offered a prize for the best idea for a mechanical harvester to be submitted in September 1843 so that a prototype could be ready for testing in the subsequent harvest. Everyone else submitted a standing machine, but Bull put forward his locomotive and a friend made a model according to his directions. The horse drawn machine moved along in the crop and a comb held the dry ears of wheat firmly while the beaters placed over the comb threshed the grain and chaff out into a box leaving the stalks in the field. This still left the winnowing of the grain and chaff to be carried out in a separate operation (Bull, 1878, p. 240). The Committee were not impressed by the radical principles of Bull's machine and declared that no machine had come up to expectations.

There it would have rested had not an agricultural machinery manufacturer (John Ridley) been so interested that he took the model and produced a prototype for testing. Bull was sent the prototype to test on his farm and he took off his standing crop with great gusto. He was particularly pleased to find that the dryness necessary for the efficient use of the machine was not the hazard or handicap that many observers believed it would be. In fact, this chip-like dryness of the grain made the South Australian wheat travel better and it became a distinct advantage. When it arrived in London dry and sound it gained the top price. When Ridley made the first machine, he interposed a modification or two onto the conformation of Bull's original design, but testing in the field caused a second manufacturer, Marshall, to revert to Bull's original design. Following the successful testing, Bull gave his model of the machine to the Chairman of the Corn Exchange Committee saying that he would forgo copyright and allow its manufacture as a service to the public. The Bull reaper was the most commonly used machine for harvesting cereals for many years. The benefit to the cereal farmer can be discerned by the fact that prior to the invention of the reaper it took one man six hours to cut 0.4 ha of grain. The reaper was capable of cutting 4 ha per day (Pike, 1957, p. 329). The cost of the reaper was over $AUS 100 per machine, and the cost of reaping by hand was around $AUS 1.50 per day plus food. The saving in cost was in addition to the security of getting the crop in rapidly and so avoiding damage from inclement weather. Although many farmers could not afford one immediately, access to the reaper increased over time until it became widespread. Ridley. who was wrongly credited with the invention and was awarded an *ex gratia* payment by the South Australian Government in recognition of this, did not copyright the invention even though he manufactured many hundreds of the machines. Later when he returned to England he built a modified version for use there, and in this case, he did apply for copyright. However, the machine did not take off, and Ridley died before he could succeed with a restructured version. Bull, on the other hand, fought for the recognition he felt was his due, and he was granted an *ex gratia* payment of £250 by the South Australian government in 1882 'because of improvements made in agricultural machinery'.

Many years later the old Collegians of Roseworthy Agricultural College decided to pay a sum to Italian sculptors in Pisa to cast a statue of Ridley to commemorate his 'invention', apparently forgetting that it was farmer Bull who had invented it. The statue was unveiled in March 1915 at the College. Unhappily the farmers of the State did not provide the support expected. Of the 103 subscription lists sent out to the Agricultural Bureaux only 24 were returned and only 14 sent money. The total raised was £15, leaving the Memorial Association with a debt of £20. 3s. 5d. The Principal of the day, Perkins, castigated the farmers in his address, claiming that they did not appreciate that Ridley had made wheatgrowing 'possible' and the College had made it 'profitable' (Bagot & Ridley, 1844–77).

Later an ex-farmer, H.V. McKay from Victoria, refined the reaper further by adding a knife to the comb so that the heads were cut and not just beaten, and incorporating a winnower into the machine that blew the chaff back into the field leaving clean grain in the box. This meant that the whole of the harvesting and cleaning of the grain could be carried out in one operation. This has continued to be the principle underlying all subsequent Australian mechanical harvesters or 'headers' as they are called. When the South Australian farmers pushed out into the marginal areas they were harvesting crops with yields of as little as 140 kg/ha and working in brittle heat. Bull's machine was ideal for these conditions and little or no grain was lost during the harvesting process. The Bull reaper set a standard for efficiency for dryland harvesting that has not been bettered.

In Europe and America the development of mechanical harvesting went down another path. There the grain and the stalk were cut off at ground level and bound into sheaves with a binder, and were then transported to a central location where they were fed into a stationary thresher and winnower. Later these operations were combined to produce a combine harvester. Even today when the modern Australian version of the McKay harvester is not much different from that of the modern European or American harvester, the standard of efficiency of harvesting of which the Australian model is capable is vastly superior under dryland conditions to that of the others, and very little grain loss occurs.

Capital investment in farm equipment

The difference in the capital investment needed for deep and shallow tillage is graphically illustrated by the contents of the machinery sheds on individual farms of similar size.

In North Africa and the Near East the implements bought and used to till the soil are designed in North America and Europe. First the farmer ploughs deep and makes great clods, then he discs and cultivates to break them down, then he prepares a seed bed and later he applies fertiliser before seeding. On a farm of, say, 500 ha one usually sees two or three crawler tractors, two or three large wheeled tractors, and two small or medium sized tractors. Tillage implements comprise four or five deep ploughs, or two or three ploughs and several chisels, the same number of tandem discs (called 'cover croppers' in North Africa), and probably some light cultivators and harrows. European or American autoheaders are used for harvesting cereals. The yield of cereals in spite of this investment in farm machinery is not high and the cost of the machinery and its operation makes the farmer dependent on subsidised prices.

On a South Australian farm with an equivalent cropping area one would find no more than two large wheeled tractors or possibly one large and another smaller model for harrowing and spraying. Tillage implements would be two large (twentyone tyne) trailed scarifiers, one light forty tyne cultivator, and one complete set of trailed harrows. Usually one of the scarifiers is incorporated in a combine seeder. A header can either be owned by the farmer himself and contracted out to harvest other farmers' crops, or the farmer brings in a contractor to harvest his crop. Sometimes a group of farmers will combine to buy a header and operate it on a cooperative basis.

Farmer research and self-help

Apart from inventing implements to overcome the physical difficulties associated with dryland farming, South Australian farmers were eager to carry out research into other aspects of farming.

As well as the Corn Exchange Committee that encouraged farmer inventions, a Royal Adelaide Horticultural Society was established in 1848. In the first few years it gave prizes and acted as a social get-together. But in 1867 its 397 subscribers began to take its role more seriously and the Society combined with farmers and manufacturers to hold trials of reapers, mowers and winnowers and at the request of the government, trials of the machines invented by Mullens, Scobie, Stott and other farmers for grubbing mallee and peppermint woods. It gave grants to two country ploughing matches. It paid for maps of agricultural districts to be made showing the climatic basis of each and provided support for chemical laboratories with facilities to carry out analyses of soils, manure and water. Three inspectors were appointed to lecture on technical matters and send in reports and recommendations to the Government. They also collected material about the red rust in wheat that became a problem in the 1860s (Thomas, 1879, p. 24).

Farmers communicated new and improved techniques to each other and formed self-help groups when droughts or marketing problems caused crises. In 1843 farmers found that cereal seed pickled in lime or brine successfully resisted the disease of smut, and a farmer named Reynell suggested that they try a bluestone/copper solution (Price, 1924, p. 160). Lime pickling remained the preferred treatment for many years until a seed merchant, Hannaford, in the 1930s found another simpler method. They selected more and better varieties of wheat (in 1865 there were fiftysix varieties of wheat being grown in South Australia) and some of these produced record crops (Dunsdorfs, 1956, p. 147). In order to deal with the problem of lack of labour as flocks of sheep on farms increased and graziers found it hard to find shepherds, farmers invented a cheap type of fencing using stretched wire and wooden posts. They used galvanised rainwater storage tanks which were light and cheap to make and erect for household and stock use in the dryland regions where every drop of rainwater had to be husbanded, as well as windmills of light construction for pumping from bores and rivers. Later windmills also were used to drive generators for households where electricity from a central supply was not available. Today, the Kondinin farmers in Western Australia carry out independent testing of farm machinery and other products on individual farms and under ordinary operating conditions, publish bulletins and booklets detailing the results, explore new techniques for both farming and farm business and rely on their own members for lively discussions and critiques of research and advice from scientific and technical institutions.

Livestock production – graziers and farmers

The production of cereals benefited rapidly from farmers' innovation, and records exist to enable the facts to be re-assembled with relative ease, but it is more difficult to chart the innovation that rescued the livestock industry. While there is quite a bit of evidence to show that many of the early farmers kept animals – dairy cows, pigs, and sheep – on their farms and sold the produce, the conventional belief is that graziers had the sheep and farmers only grew wheat and barley. Certainly this is the impression given by most historians writing of the period. There are references to a lack of manure being available for crops because the few animals were allowed to wander around all day instead of being penned (Williams, 1974, p. 269). Further on we will examine farmer evidence which suggests that many farmers did keep sheep for commercial purposes even in the early days. What is certain is that the colony's graziers were treated and behaved differently to the farmer. The grazier came in with a bag of money, bought his land privately and at less than the farmer paid, put a manager on his land, then retired to the city where he lived well on the profits. The large houses built in Adelaide by those early graziers tend to confirm this. However, the low prices for wool and the lack of a prosperous meat market lead one to believe that much of the high city living must have come from other funds such as investments in the mining industry that grew out of the rich copper deposits found in Kapunda in 1843, Burra in 1844 and Kadina in 1863 (Williams, 1974, p. 134).

This influx of wealth would have come at a critical time for the early graziers, as by 1860 the natural pastures in the rangeland had been eaten out and the livestock production was falling.

The Commission controlled all land sales in the colony and it decreed that the arable land within forty miles of the capital, Adelaide, was primarily to become a farming settlement. In addition, Special Surveys of large tracts of land further out (up to 8000 ha apiece), were sold off by private treaty to those who could put down $AUS 16 000 in cash either in London or in Adelaide. Pastoral leases for grazing were also available and the conditions governing them were complex and changed from time to time as political pressure from graziers or farmers in turn gained dominance. In 1846 the colony had less than half a million sheep and licensed stockholders could run sheep and cattle outside alienated land by paying $AUS 0.05 per head for cattle and $AUS 0.01 per head for sheep. The only boundaries they had to recognise were those that signified the Hundreds, the name for the collection of surveys that made up the alienated surveyed land available for purchase. In addition, annual leases of unsold land within the Hundreds were rented to licensed stockholders when they needed to bring their livestock in because of drought. The common-land pastures within the Hundreds (to which all freehold landowners had access) were also available to the stockholder who was given a quota. In 1850 this changed. This quota system was abolished, and the stockholder had to negotiate a lease at $AUS 5.20/sq. km if he wished to continue grazing within the Hundreds. The leases were for sections of 260 ha so as not to discourage grazing. Outside of the Hundreds, fourteen leases were made available at rentals that reflected the quality of the pasture available (at $AUS 5.20, 3.90 and 2.60/sq. km) and the boundaries of the leased territory had to be carefully defined before the lease was signed. This meant a degree of security of tenure for the stockholder, but land close to the Hundreds was liable (after due notice) to be reclaimed by the Government and proclaimed as Hundreds for subdivision and sale. In 1846 a Control of Wastelands Act was passed in order to stop the granting of more Special Surveys because of the hostility that had developed between the grazier who had access to large areas of cheap land at the preferential rate of $5.20/sq. km and the farmer who had to pay $4.80/ha for his section. The Special Surveys were sold by private treaty and not at public auction as were the farming sections, and this exacerbated the hostility. Even when the Special Surveys were stopped, most graziers ran their sheep on cheaply acquired leases adjacent to freehold sections.

By 1851 the flock had increased to 1 250 000 but the natural pasture was beginning to show signs of stress and as it became eaten out in the better areas, graziers were forced to take up leases along the upper reaches of the Murray River and so on into the more arid north of the State (Pike, 1957, pp. 324–8). Bull refers in 1846 to the whole length of the Murray River from North Bend to the coast 'lined with sheep and cattle stations' (Bull, 1878, p. 180). Prices for wool went up and down due to various influences, drought, disease, export demand, and so on. The concentration was for the most part on wool production, using flocks that originated from the Australian Merino sheep introduced from Spain by John and Helen Macarthur in New South Wales (Bull, 1878, pp. 24, 38, 146, 174). These were driven overland to South Australia from Victoria and New South Wales and were preferred to those available from the island of Van Diemen's Land offshore, as these sheep introduced scab to the colony and imports had to be stopped and a campaign undertaken to treat the disease (Pike, 1957, p. 326). Meat was sold to local markets, particularly when wool prices were down and when things got really bad, sheep were boiled down for tallow. By 1850 Saxon rams were being imported from Germany to improve the flock and the average wool clip per sheep rose from 3.5 lb (1.6 kg) per year in 1850 to 4.75 lb (2.15 kg) seven years later. However, the lack of pasture continued to constrain the growth of the flock. In 1845 wool had made up 55% of the State's exportable product; by 1851 it had dropped to 25%. In 1897 it was claimed that there were more sheep being kept on farms than on pastoral leases and the estimate for sheep numbers was seven million (JAI, 1897, p. 548).

By 1902 the state flock, as a result of a devastating drought, declined to four million (SAYB, 1970, p. 403). While it is difficult to be absolutely sure, it is likely that the number of sheep increasingly being kept on farms was disguising the rapid decline in flocks being held in the rangeland as the pasture was eaten out. Certainly the graziers who had land in the high rainfall country of the Mount Lofty Ranges, the Barossa Ranges, and the South East still had good natural pasture, but they were unable to increase their flocks because if they did they would lose their natural pasture. There was hostility between graziers and farmers. Graziers were seen by the farmers as jumped-up pretenders to an aristocratic position based on the grazier's snobbish attitude to those who tilled the ground. Farmers believed the only work the graziers did was to go to the city office of woolbuyer Elder to collect their periodic wool cheques. The graziers looked down on the farmers – mucking about in pig pens and dairy sheds and digging and toiling with their hands. The social distinction between the wife of the grazier who lived in the city and never touched the land, and the farmer's wife who carried out all manner of tasks in order to keep the land and its produce profitable, began at the same time. Raouf Sa'd Abujaber relates a similar superior attitude among Bedouin nomad tribes towards sedentary farmers in Jordan (Sa'd Abujaber, 1989, pp. 177–97). Both attitudes still lingered in South Australia until the mid-1970s, when the graziers reluctantly combined with the farmers to create a single organisation in order to become more effective in lobbying governments for legislative privileges and access to government funds.

For the purposes of this book, however, we must put aside an interest in this curious social attitude and confine ourselves to considering the consequences of the situation at the end of the nineteenth century when in South Australia there was a decline in the numbers of sheep in the arid zone because overgrazing had destroyed the ability of the land to regenerate ephemerals. The saltbush and bluebush that remained were not capable of providing more than a drought maintenance diet for sheep.

Cereal production

The answer would eventually lie in a change in the method of cereal production.

The first wheat crop was exported from Adelaide in 1842, (Forster, 1866, p. 329). By 1880 wheat was the dominant crop and South Australia, in spite of many periods of low prices, was the major Australian exporter (Dunsdorfs, 1956, p. 113). Farmers utilised the efficiencies and cheapness of shallow cultivation, mechanical harvesting and cheap land clearing techniques to sow wheat to the exclusion of all else. In his excellent geographical study of the South Australian landscape, Michael Williams suggests that the mechanisation of land clearing, shallow cultivation and harvesting also encouraged the early farmer to extend his farming out into the scrublands and eventually to what we now consider to be the limits of arable land in South Australia. The allowable average size of cereal farms was increased by government regulation, partly due to the administrative difficulties of controlling holdings at a limit of 32 ha per farmer, and partly as a result of a combination of the efficiencies of mechanised operations and the demands of farmers for more land to offset falling yields. By 1860 holdings of 130 ha and 260 ha were common, and by the 1880s many cereal farms were 400 ha or more (Williams, 1974, p. 115). Yet in spite of the successful farmer innovations which mechanised cereal production, farmers research into better wheat varieties and the expansion of individual holdings, cereal yields that began to fall in the 1860s continued to do so, even in good years and in good areas, and by the 1880s were as low as 600 kg/ha. Certainly some of the drop in averages must be laid at the door of expansion into what was considered to be marginal country – where average rainfall is between 200 and 250 mm and where crops were grown without fertiliser – but this was not the only reason. In 1903 practically the whole of the wheat crop was grown in three districts – the Central with a rainfall of 530 mm sowed 220 958 ha; Lower North with 430 mm sowed 216 639 ha, and Upper North with 350 mm sowed 195 265 ha (Gordon, 1908, p. 106).

As early as 1841 Bull warned that although there were crops on all sides of the city and wheat was yielding from 2100 to 2800 kg/ha, there could only be a 'diminishing in quantity of yield through exhaustion and bad management' (Bull, 1878, p. 53). He was afraid that the economic pressures of low prices and high costs would force constant cropping and that this would lead to soil exhaustion. He suggested that the government should include in its conditions of land purchase a regulation that no more than one cereal crop every three years should be allowed. He believed that a three year program similar to that known as the Northampton or Bedford rotation should be adopted. This would result in wheat being grown on sheepfold land, with grazing one year in three, and one year fallow. His folk memory of Northampton was that this rotation had been in operation over 'many hundred of years' and that such a rotation would be vital in such a dry climate as South Australia. He also considered the mandatory holdings of 32 ha to be insufficient for such a rotation because of low prices for livestock products and cereals (Bull, 1878, p. 56). Bull was at pains to point out in his overall consideration of the wheat industry in South Australia that the first crops had come from good virgin soil; but the second from a large proportion of second and third rate land – 'much of it reduced in productiveness after years of the usual colonial exhausting courses of wheat after wheat, as long as a fractional yield can be got'. He was wrong about costs being the spur to thrashing the land. His own invention of the reaper and those of Stott, Smith and Mullens had dealt with this, but prices were never very encouraging. In the 1880s prices remained as low as ever, and the response of the farmers was what Edgar Dunsdorfs later called 'the perverse reaction' to prices, that of expanding area in response to low prices in order to recoup their losses (Williams, 1974, p. 155). Even today farmers tend to expand their cereal area after a drought in order to recoup their losses (Chatterton & Chatterton, 1981, pp. 147–63). Bull's opinion that farmers needed more land in order to farm better was not vindicated when the Strangways Reform in 1869 gave farmers the opportunity to buy larger farms on credit. It simply meant that they went out into the drier areas and used the same methods.

One must not be too hard on the farmers though. They were very concerned at the problem and although most of them produced a wheat crop most years, most took care to rest part of their land by incorporating a year of rest fallow into their wheat rotation (Dunsdorfs, 1956, p. 138). Kyffin Thomas provides further proof of this when he described the common practice around the state in 1879:

The system of wheatgrowing in the colony is extremely simple. The selector merely ploughs the land to a depth of a few inches, seldom more than 6, the seed is generally sown broadcast, and generally, if it is not put in after the proper season, and if there is no visitation of red rust or other destructive plaque, a crop of from 10–20 bushells per acre is obtained, according to the quality of the land. In the older agricultural districts where yield is often small as there are difficulties to be coped with in the shape of weeds and exhaustion of soil . . . the commonest plan in most cases is to allow the land to lie fallow and depasture cattle upon it for a year or two after which it will again yield a renumerative crop of wheat. The keeping of sheep has of late years become a favorite expedient for recruiting the soil (Thomas, 1879, p. 20).

Yet while many armchair observers and many farmers believed that a rotation which enabled the farmer to produce some other crop as well as wheat was desirable no one seemed to be able to suggest anything that was possible within the climatic and economic resources available to the farmer.

Science comes to South Australian farming

In 1875 the South Australian Government, at the urging of politician and grazier Sir Henry Ayers (later Premier of the State) and the armchair critic Albert Molineux, appointed an Agricultural and Technical Commission to investigate the need for a structure to provide a scientific input into farming. In the same year Molineux (a self-educated printer) began to publish *Garden and Field*, in which he continually expressed great scorn for the South Australian farmer and contempt for shallow cultivation – these farmers who scratched the soil and farmed sloppily. He and his friends undertook a long campaign urging the colony's farmers to duplicate the English farmer with his root crops, dairying, vegetable and fruit on all farms, deep ploughing and guano for the land and hedgerows for the fields. Their cries of 'We told you so' intensified as they noted the continuing decline of wheat yields and the signs of erosion that were appearing, particularly in the marginal areas (Black & Craig, 1978, p. 24).

The Commission filed a recommendation for the establishment of an agricultural college at Roseworthy in the mid-North of the state, together with research farms, one attached to the college, where trials and then demonstrations of improved farm operations could take place. The college was to train the minds of those who would be involved in farm advisory work, to conduct scientific experiments and to set an example by being ordered and disciplined in the approach to farming activities. A Department of Agriculture was to be established to formulate regulations and provide inspections to underpin the agricultural infrastructure. The Commission clearly saw science as providing an intellectual resource – a partner in agriculture, not the authoritarian body it eventually became.

Farmers were outraged that the Government should doubt their competence to solve their problem. They resisted the setting up of the Commission and when it made its report they resisted the Government's decision to bring them the benefits of what Europe was calling 'the "Golden Age" of British agriculture' (Callaghan & Millington, 1956, p. 2).

However, another four years of poor wheat harvests stiffened the resolve of the Government, and in 1881 J.D. Custance was appointed the first Principal of Roseworthy Agricultural College. He came from England and took up his post when the college was opened in 1882. In looking to England for a suitable Principal for the new college, the S.A. Agent General had been asked to find the best man for teaching 'a more rational system of agriculture' (Molineux, 1887, p. 102). Custance had trained in England and then gone to Japan where he advised the Japanese Government on agriculture and established for them a college of agriculture. He lasted five years in South Australia, not because his training made him unfitted for the position, but because of his abrasive attitude to the farmers whom he was engaged to influence and teach. He was too authoritarian for the farmers and the Commission's vision of a scientific partner for the practical farmer remained elusive. The patronising attitude adopted towards the farmers is illustrated in the account given by Allan Callaghan in 1956 of the early role of the South Australian Department of Agriculture. Its first contribution in the 1880s was to get together a staff of 'practical experts' who provided advice to 'settlers who knew little of horses or machinery and even less of agriculture' (Callaghan & Millington, 1956, p. 445). It is little wonder that the farmers were irritated.

The first shots Custance put across the bows of the South Australian farmer were that they must be like English farmers, adopt crop rotations, use fertiliser and plough deep.

Farmers were exasperated that Custance should tell them that they should adopt a farming system they had already found to be inappropriate for the climate in which they lived and farmed. They had tried and failed to grow turnips and perennial clovers and other rotations dependent on high rainfall and ample labour. Many were keen to use fertiliser but guano had proved too expensive, animals were grazed extensively and could not provide the folds full of manure that the English farmer utilised, and they were quite happy with their shallow cultivation because it was cheap, easy and effective.

Shallow cultivation vs. deep ploughing

The battle to change from shallow cultivation to deep ploughing was over before it began.

Custance, appalled at the apparent slovenliness of shallow cultivation, immediately stomped the countryside calling on farmers to abandon it and take up deep ploughing.

The farmers refused.

He told farmers that the header and the stump jump plough 'had no future' and the farmers shook their heads at his ignorance. The editor of the *Gawler Standard* in 1882 responded on behalf of farmers that 'Practical men know it is madness on the . . . soil chiefly found in South Australia, to deep plough' (Molineux, 1882, p. 26).

Neither side would give in.

Custance's obsession with deep ploughing was carried on by his successor Lowrie and then by successive Principals of Roseworthy and their colleagues for many years. When A.J. Perkins retired as Principal in 1930s he made a last, not very convincing, plea for deep ploughing (the mouldboard and inversion of the soil) – not the deep ploughing that Custance had proposed, but deeper than the 10 cm the farmers stubbornly refused to discard (Perkins, 1934, pp. 696–717). There was no response.

No early technical expert appears to have considered the importance of the savings in cost and time to the farmer who used shallow cultivation and rejected deep ploughing although this is a major factor in the farmer's adherence to the practice. Research as to any difference in yield following the use of each style of cultivation was favourable to shallow cultivation – particularly if the correctly designed implement was used for the operation. (Spafford, 1927, p. 426).

In the event, the farmer's faith in shallow cultivation in dryland farming conditions has been vindicated and the challenge today is to convince other farmers (not so fortunate as to have escaped the deep ploughing phase) that their crops will not suffer from shallow cultivation and that their land will benefit.

Conclusion

In order to see the great differences that came into being in the latter part of the nineteenth century between the Australian dryland farmer and his counterpart in the Northern Hemisphere one must recall the leaps in knowledge that took place at that time.

While the South Australian cereal farmer was learning to use mechanised shallow cultivation to till the soil, control weeds and prepare a seed bed in the rapid way demanded by dryland conditions, to select wheat varieties for disease resistance and climate, and to mechanically harvest in such a way that the effect of the dry atmosphere of the crop became an asset, the English and French cereal farmer was learning to cope better with a humid climate – to plough deep, to plant root crops and perennial clovers in rotation, to substitute nitrogen fertiliser for organic matter, and to use a long ploughed fallow as a means of eliminating weeds and spreading the load of working the land over a longer period.

In the Near East in dryland conditions similar to those of South Australia, cereal farmers were using wooden ploughs drawn by mules to plough the land (not very deep) and continued to broadcast seed by hand. Fertiliser was unheard of. Fallow consisted of leaving the land to lie covered by weeds and grasses and annual legumes for a year or two while another piece of land was ploughed. Animals were grazed on most farms but were fed with hay in winter and grazed on pastoral lands in spring. In late summer they were brought back to the farm and ate the cereal stubble until late autumn. In addition, large flocks of sheep, camels and goats were kept by graziers in the rangeland and taken to the cereal zone for stubble grazing in the summer. When there was a drought, many livestock died. When the season was good they multiplied again (Sa'd Abujaber, 1989).

In North Africa, the French influence brought in by colonists in the nineteenth century meant that on the better farm land the European system was used and deep ploughing (up to 60 cm) became common. The French tended to use animal manure for their crops (mainly from cows kept in biers) and the Arab farmer on the small holdings in the better areas, in Morocco at any rate, followed suit. A long ploughed fallow to destroy weeds, a long, tortuous preparation of the seed bed and nitrogen fertiliser were commonly used on large farms at the end of the nineteenth century. In the semi-arid zone where most of the Arab farmers had been pushed, the pattern was similar to that of the Near East.

In the twentieth century, particularly after the 1950's when the aid and development agencies (spearheaded by FAO) concentrated efforts on mechanising and modernising the cereal and livestock farming systems of North Africa and the Near East, the larger Arab farmers came more and more under the influence of the Northern European system of cereal production and started on the long trajectory that resulted in the farming system they practise today. The smaller farmers, in spite of their role as justification of the many aid programs, remain for the most part as they were.

This adoption in North Africa and the Near East of nineteenth century discoveries appropriate to cereal farming and livestock production in England and France and later North America, led in the latter part of the twentieth century to soil infertility and erosion of prodigious proportions. Much of the region today is an example of what happened in South Australia when, towards the end of the nineteenth century, natural pasture in the rangeland had been largely eaten out and erosion of farmland was conspicuous and spreading in spite of efficient tillage and harvesting. At that period the search for a rotation within the means and climatic restraints of the South Australian farmer became the focus of farmer and technical expert alike. What worked in Europe could be tinkered with, but it seemed as though something entirely new may have to be found if the soil was to become, and remain, fertile.

3
Medics and sub-clover on the farms

Introduction

The story of how farmers discovered and then stuck to their own way of using medic and sub-clover in spite of strong advice to change reveals fundamental differences in the approach taken by technical experts and farmers when tackling farming problems. Both groups began with a common goal. They wanted to find a source of feed for sheep other than the native pasture and bushes that could not stand the pressure of commercial flocks and they wanted to stop the erosion and reverse the cycle of diminishing soil fertility that threatened to destroy the state's wheat industry. The State Government endowed the scientific and technical community with funds and land to enable them to concentrate on these problems. Farmers got together at meetings and after work to share their experience.

Sub-clover pasture for livestock production

In 1887 a farmer and nurseryman, Amos Howard, noticed a strong growth of annual sub-clover (*T. subterraneum*) in his paddocks in the high rainfall (600 mm) Mount Lofty Ranges east of Adelaide, and he began to graze his sheep on it. His sheep did well, and the pasture regenerated each spring without further seeding. He told his neighbours about it. In 1900 he harvested the first sub-clover seed (calling it Mount Barker in honour of its location) and in 1903 he was selling commercial quantities of seed at $AUS 1.50/kg. In 1905 he invented a method of 'podding' or threshing the pods to extract the seed easily. In 1906 he wrote to the newspapers describing the way he and his neighbours were profiting from the grazing of sub-clover (Williams, 1974, pp. 313–14, Donald, 1982, pp. 61–2).

Other graziers followed his example and in his own region sub-clover became a common pasture. In the high rainfall Lower South East (between 500 and 600 mm) where large grazing leases existed, graziers running cattle and sheep showed an interest in sowing sub-clover pasture. In 1901 a farmer found after putting superphosphate out with his cereal seed that it improved the quantity of clover in his paddocks the next year. He asked 'If the stocking rate of sheep is doubled by grazing local clovers, why not put superphosphate directly onto the pasture?' (JAI, 1901, p. 731). When farmers followed his example and their sub-clover was dressed with superphosphate at a rate of around 100 kg/ha, it dominated paddocks and provided excellent sheep feed. It provided green feed in the cool months and dry feed from the pods and straw in the hot months. It was cheap and once established it did not need re-sowing. There was no need to 'improve' it with English or South African grasses or to grow fodder crops such as oats, barley, or kale and evening primrose to provide extra feed. This was the advice being given by the Government agencies.

W.L. Summers, a fertiliser inspector who travelled large areas of the state's farming and grazing land, provides evidence about the use of sub-clovers by farmers and graziers at the beginning of the twentieth century. Each time he presented his annual report of fertiliser usage in the *Journal of Agriculture* he managed to include some more news about the ubiquitous nature of sub-clovers and annual medics on grazing properties and farms and the valuable use to which farmers were putting them.

In 1901 he was reporting that many farmers said that the increased productivity from the abundant grazing available for sheep easily repaid the cost of applying superphosphate to the pasture. He believed that this better quality pasture was improving stock health and asked for experiments on its potential to be carried out by technical experts (Summers, 1901, p. 810). In 1907 he wrote that 'sub-clover is probably the most valuable of our naturalised fodder plants in the cooler districts'. He reported that farmers and graziers were finding that it 'withstands hard grazing'. He also reported that various newly identified medics accidentally introduced onto farmland were, after the addition of superphosphate, colonising rapidly and producing 'excellent pasture plants' (Summers, 1907, pp. 375–8, 577–9).

An Angaston farmer reporting back to his branch in 1909 said that he saw 'clovers growing luxuriantly, up to three or four feet high' on a farm owned by Smith, a farmer on Kangaroo Island, who was getting excellent grazing from them for his sheep (JDA, 1909, p. 449).

The reports of these excellent clover pastures were not unnoted at Roseworthy. In 1911 A.J. Perkins told a meeting

Specially noteworthy too has been the free development of leguminous plants, so much so indeed, that clovers appear to have sprung up spontaneously in fields in which hitherto their presence has not even been suspected. There is no doubt that this increased stock carrying capacity of our farm lands must in the end react favorably on their general fertility and crop carrying capacity . . . (Perkins, 1911, pp. 10–14).

Yet this apparent enthusiasm did not have much effect on the interest of Roseworthy in the behaviour or feed potential of these adventitious clovers and medics. Research into forage crops and fodder conservation continued to be based upon introduced grasses and crops.

In 1922 farmers in Kybybolite district in the South East of South Australia were reported to have sown large areas of sub-clover pasture at the rate of 560 g to 1.12 kg/ha and advertisements in local newspapers for Howard's Mt Barker sub-clover seed were common (JDA, 1927a, p. 61).

It was not until 1924 that the Department of Agriculture finally set up a trial at the Kybybolite Research Station where L.J. Cook sowed one quarter of an acre of sub-clover and reported an 'excellent result' (Cook, 1927, pp. 36–40). The Department went on to duplicate the experience already pioneered and well known by farmers and graziers.

In 1927 farmers were writing of profitable lamb and wool enterprises from sheep grazed on medic and sub-clover pastures to which superphosphate had been applied. Some farmers had tried sub-clover on Western Eyre Peninsula but the conditions were too dry (JDA, 1927b, pp. 604–7).

It was not until 1930 that the Department tried to grow sub-clover at Saddleworth in the mid-North and on Western Eyre Peninsula – both dry regions – and, just as the farmers had found and reported, the trials were not successful.

Farmers in areas where rainfall was below 400 mm annually but where the acidity of the soil favours sub-clover, were later able to use the cultivar Dwalgenup which had been discovered and identified by a farmer in Western Australia.

The effect of sub-clover on the industry

In 1880 the natural pasture in South Australia, particularly in the rangeland, had been under great stress and in parts completely eaten out. This had caused a stop to the initial burgeoning of the state's flock. The major part of the arid rangeland did not recover from this initial shock. Grazing properties in the high rainfall districts were constrained from increasing their flocks by the threat of losing the natural pasture. Sheep production only expanded again after sub-clover provided reliable and high quality grazing in this high rainfall country. What attracted graziers and farmers to sub-clover was that it cost so little to establish and to maintain. Once sown or found sufficiently colonised, it required only dressings of superphosphate and sensible grazing to keep it in good condition. Farmers and graziers seemed to have had no trouble in judging how to graze it to ensure that it retained a sufficient seed reserve for successful and adequate regeneration. It does not appear that farmers ever saw it as a pasture that should be re-sown. Seed was bought to establish new areas, not to re-sow existing ones. Constant grazing seemed to maintain a clover-dominant pasture although grasses and other volunteer plants were a part of the pasture.

However, the real explosion in flock size did not occur until the 1930s when the use of medic pasture to replace bare fallow on cereal farms became widespread.

Medic for livestock on cereal farms

Reports of early medics colonising on farm land are contemporary with those of sub-clover. At about the same time that Howard noticed and began to encourage sub-clover pasture in the high rainfall zone, some farmers in the drier areas on the plains and on Yorke Peninsula noticed medics growing strongly in paddocks in the rest fallow before barley was sown. Other farmers noticed it in the rest fallow following a wheat crop. They began to increase the number of sheep on their properties and found that by balancing the number of sheep with the amount of pasture and cereal stubble available they rarely needed any other fodder to supplement the flock's requirements. Farmers were quick to take advantage of these volunteer medics (*M. minima* or *M. polymorpha*), and in 1909 a farmer from Wilmington (in the mid-North) wrote that 'generally most farmers keep 200 sheep'. The Wirrabarra Agricultural Bureau spent an evening discussing the fact that '10 years ago few farmers kept sheep, now most look for the annual lamb cheque' (JDA, 1910, p. 621). By 1897 light, cheap wire fences had been adopted by farmers to keep horses and other livestock in paddocks and shepherds had become a thing of the past. A farmer had wondered why other farmers did not follow his example in using these fences to hold flocks of sheep from which to sell meat and wool (JAI, 1897, p. 63).

In 1911 the Principal of Roseworthy College noted that the 'common enough rotation north of Adelaide was bare fallow/wheat/pasture'; however, he told farmers that 'one need not rely on adventitious weeds' – then widely used as a synonym for annual medic by technical advisers – 'but can oversow vetch or drill-sow lucerne and then harrow' (Perkins, 1911, pp. 12–13). As farmers were quick to point out, lucerne was not appropriate for a pasture/cereal rotation because it would not survive the next fallow and cereal crop, and vetch required re-sowing and haymaking and thus increased the work load. The labour involved, the cost of seed and tillage, and the moisture requirement for these more elaborate programs were beyond the capacities of the South Australian dryland wheat farmer. He continued to find his answer to sheep feed in 'adventitious' pasture.

In the same year that Roseworthy was suggesting that volunteer medic pasture could do with improvement by oversowing with vetch or lucerne, farmer members of the Nantawarra Agricultural Bureau discussed whether it was more profitable to run a fallow/wheat/pasture rotation or to continue a bare fallow/wheat rotation. 'Most farmers,' they agreed, 'have a lamb crop' and use a 'three year rotation of fallow/wheat/pasture'. One farmer said he made more out of growing wheat for seven years, another said he did better out of lambs over the same period and a third said a combination of each was the best solution but it was agreed that keeping sheep did not pay if you had to grow and make fodder to feed them (JDA, 1912, p. 173).

In 1935 A.B. Ferguson, a farmer at Arthurton, told a conference of farmers on Yorke Peninsula how his pastures grew more abundantly after he had applied superphosphate fertiliser to medic growing on his farm. He expressed doubts about the value of Wimmera ryegrass (much promoted by the Department of Agriculture at that time for pasture improvement) which he considered a 'weed' (Ferguson, 1935, p. 1069).

An 'unsuitable' cultivar

One of the first medic cultivars that colonised on farms in the state was *M. minima*, the pods of which are covered with burrs. As early as 1903 farmers were reporting that while the 'native clovers appear after super' the burrs were 'objectionable in the wool' (Foster, 1903, pp. 98–9). In 1927 a farmer (J. Fradd from Beetaloo Valley) mentioned the good growth of medics and grasses on his farm but said that he had to accept a price cut for his wool because of the medic burrs that contaminated the fleece. The penalty was compensated for by the increased number of sheep he could carry because of the medic pasture (JDA, 1927, p. 1185).

In South Australia, unlike North Africa and the Near East where medic is indigenous, if an adventitious variety is found in the paddock and it causes a problem, a more suitable cultivar can be sown and this will change the composition of the pasture, and this is what happened in this case. A burrless variety of annual medic (*M. truncatula*) had been identified on farm land in South Australia as early as 1909. It was noted by the Waite Institute in 1925, and seen again at Gepps Cross near Adelaide in 1931. It was seen again at Saddleworth and in the Mallee in 1932 and its existence and characteristics were published by Dr H.C. Trumble from the Waite Institute in 1937 (Trumble, 1938, pp. 953–8).

Trumble's paper listing the characteristics of several varieties of medic, among which was the burrless *M. truncatula*, was seen by a seed merchant, ex-farmer, Alf Hannaford. A typed draft of the paper was found in the 'Hannaford papers'. He produced some seed from this cultivar on a farm at Port Noarlunga in 1938. He used this seed (which he named 'Barrel medic' or 'Barrel clover') as the basis of a publicity campaign contrived to make farmers aware of its existence and benefits. He wrote newspaper columns, paid for advertisements to be published, and from a stand at the Royal Adelaide Show he distributed pamphlets telling farmers that this burrless medic would increase soil fertility for the cereal crop, reduce the risk of erosion and provide grazing without adulterating the wool of the flock. Due to the large response he got from this campaign he had to look around for more seed. He obtained it from the process of cleaning cereal seed. At this period herbicides were not used to control weeds in crops and so some of the natural medic and sub-clover germinated during the cropping phase. The resulting medic and sub-clover pods were a contaminant in the cereal grain and Hannaford's company was employed on contract to clean the grain. From among the 'rubbish' rejected from the grain, the pods were separated, cleaned, and the seed extracted by threshing. The seed was then scarified and sold to those farmers who wished to extend or establish medic pastures on their farms with 'Barrel' medic.

In 1942 (six years after a solution had been found) S. Williams, on behalf of the Department of Agriculture, carried out trials that showed that burrs in wool reduced the price received by the farmer by one sixth (Williams, 1942, p. 45).

Farmers had already made these calculations in 1927 and decided that even when their sheep grazed medic full of burrs, the increased production from the farm made possible by the pasture more than made up for the small penalty incurred. Hannaford's 'Barrel' medic gave them the means to overcome this penalty long before the Department of Agriculture had carried out its research.

Medics on the driest farms

By 1938 medics were producing good pastures even in the driest and most marginal of South Australia's wheatbelt, and *M. truncatula* was noted as being 'an outstanding plant on farms in the Mallee' (Trumble, 1938, pp. 953–8). The Mallee is, with Western Eyre Peninsula, the driest (200–250 mm) and most vulnerable to erosion of all the South Australian wheatland yet sheep are the major farm enterprise now and cereals are only sown when rains are good.

The effects of medic on the industry

As stocking rates increased, farmers carefully balanced the number of sheep per paddock to ensure that overgrazing did not deplete the seed bank and endanger the natural regeneration of the pasture.

By the 1940s the self-generating medic pastures in the dry regions and the sub-clover pastures in the higher rainfall regions became solidly established as the basis for livestock production in the State. Grazing in the arid rangeland became an anachronism. By 1969 of the 18 300 000 sheep in the state only 1 200 000 were kept in the arid rangeland (SAYB, 1970, p. 404).

The rapidity with which legume pastures enabled farmers to restock after severe drought is illustrated by the fact that in 1977–8 (a period of abnormal drought) sheep numbers fell to their lowest level for twenty years. They steadily recovered to reach over 16 million in 1983–4 in spite of setbacks of further drought and bushfires in 1982–3 (SAYB, 1985, p. 386)

Medic for livestock on research farms

Dr Callaghan, when he was Principal of Roseworthy, is credited with recognising 'the potential of deliberately integrating a medic legume-ley for fat lamb production into the cereal cropping program' (Spurling, 1987, 61–7).

He did not, however, follow the farmers' method and on his Roseworthy College farm he had the flock fed with fodder made from oats, barley, kale, English grasses, lucerne and perennial clovers. Hay was cut from the medic and grass pasture. Fed on this conserved, labour-intensive fodder (oats and lucerne were the favourites), the Roseworthy College flock increased from 939 sheep in 1933/4 to 1131 in 1938/9 – an increase of 20% (JDA, 1939).

In 1934 farmers in the neighbourhood were reporting that they had increased their stocking rates from one sheep per three acres to three sheep per one acre by grazing them on regenerating sub-clover and medic pastures all year round and dry cereal stubble when it became available in the summer. This was an increase of 900% (Wray, 1934, pp. 1560–61).

On the research farms, however, regeneration of the pastures was not the norm. Their objective was to increase sheep production using conserved fodder and so the medic was cut for hay and, when this was followed by a season of bare fallow, the seed bank was excessively depleted and re-sowing became necessary. Until the mid-1950s, Minippa Research Centre farm and Roseworthy College farm regularly set aside a section for seed production in order to re-sow their medic pasture. The more lavish resources available to the research farms, combined with the orientation of the technical experts towards the model of European livestock production systems, caused them to ignore the cheap alternative preferred by the farmer who could not afford to adopt the labour intensive regime required for haymaking and silage or the time for feeding out, let alone the seed bed preparation and re-sowing necessary to re-establish the pasture each year. It was not until 1962 that official recognition was made that 'Hay isn't the only way' and the technical establishment fully endorsed the farmers' simple and practical exploitation of regenerating medic and sub-clover pastures (Pearson, 1962, pp. 23–7). The scientific attitude to medic and sub-clover was that it was an 'adventitious' component in rest fallow. Little if any research was done into legume pastures in the early days. Wheat was king to the scientific community and when superphosphate began to be used for the crops its effect on the subsequent volunteer medic pasture was regarded as incidental. The economic benefit to the farmer of regeneration of the pasture was also either discounted or ignored.

Improving soil fertility on cereal farms

Wheat production was in jeopardy by the 1880s due to soil erosion and low soil fertility, and the first task of the newly appointed Principal and staff of Roseworthy College was to try and find a system that would, above all, restore wheat yields to their early level. The value of the medic pasture to the cereal crop was very slow in becoming apparent.

The College staff drew up a three point plan to deal with the problem of falling wheat yields based on fertiliser, a change from shallow cultivation to deep ploughing, and the replacement of the common rest or weedy fallow period with a cultivated bare fallow. They were also keen to find a suitable crop or enterprise that could extend the rotation of cereal/bare fallow in order to help maintain soil fertility as well as providing some additional farm income.

The first challenge was to find a fertiliser that was suitable for the dryland conditions of South Australia.

Superphosphate fertiliser

Lawes at Rothamstead had used superphosphate in 1857 successfully to increase wheat yields and Custance the newly appointed Principal at Roseworthy believed it might prove suitable in South Australia (Callaghan & Millington, 1956, p. 91). Between 1882 and 1898 Custance carried out trials at Roseworthy using superphosphate, nitrate and guano. Initial trials with guano, which was only available in small and expensive quantities, were promising but trials were discontinued, Custance deciding, as farmers had before him, that its use was impractical (Molineux, 1886, p. 102).

The year 1884 was good, with rainfall of 543 mm and trials at Roseworthy demonstrated that the addition of 250 kg/ha of superphosphate returned a yield of 1400 kg/ha of wheat. A trial of 250 kg/ha of nitrate yielded 1412 kg/ha and plots without any fertiliser yielded 1260 kg/ha. The next year, 1885, was dry, with only 304 mm of rainfall and the application of superphosphate was increased to 375 kg/ha. The result was a yield of 1540 kg/ha. The nitrate or sulphate of ammonia did not produce a good response and even at the heavy rate of 250 kg/ha only yielded 812 kg/ha. Plots where no fertiliser was used yielded 882 kg/ha (Molineux, 1886, p. 102).

Convinced of the value of superphosphate in increasing wheat yields, and also convinced that an exponential application would result in an exponential increase in yield, Custance stumped the countryside urging farmers to buy and apply 187 kg of nitrate of soda and 625 kg/ha of superphosphate (or 'super' as it soon became known in a country where laconic speech then and still truncates words of more than two syllables) (Williams, 1974, p. 283).

Custance was not totally convinced that nitrogen was useful in the dry and capricious seasonal conditions in the state and he did not continue to press for the adoption of this fertiliser. This was fortunate as the use of nitrogen fertiliser tends to depress the growth of legumes and had farmers adopted it the medic revolution may have been stopped in its tracks.

In spite of the success with superphosphate at Roseworthy, few South Australian farmers battling with wheatsick land took up Custance's advice. In 1896, after nine years of constant persuasion and importuning of farmers (not only by Custance but also by his successor Lowrie), only $AUS 5540 worth of superphosphate was used on farms.

Costs of application

The reasons farmers gave for not taking the advice were that, in the first place, they could not afford to apply superphosphate at a rate of 625 kg/ha. The increase in yield did not compensate for the extra cost, which worked out at $AUS 19.75/ha. Wheat prices were too low to allow for such a cost. In addition, the application of superphosphate was slow and difficult. Farmers either had to broadcast it by hand, an operation that was time consuming when farms were 260 ha and more in area, or put it out with a spinner – a rudimentary implement that was profligate in its demand for seed and super. In the eyes of the farmers Custance's results reflected not a 'rational approach to farming' but the lavishness of the research farm that had labour and resources to spare (Williams, 1974, p. 282). Custance fulminated at their stubbornness, and the farmers' irritation with his attitude became so widespread that in 1887 he gave six months notice and resigned (Molineux, 1887, p. 102; Williams, 1974, p. 281).

It was a farmer, Joseph Correll (from Minlaton on Yorke Peninsula), who found a way by which super could be used efficiently and successfully in dryland conditions in South Australia. In 1896 he presented a paper to the Eight Congress of the Agricultural Bureau of South Australia detailing the experiments he and neighbours, Nankivell and Cudmore, had carried out using a seed drill to place cereal seed and superphosphate together in a shallow seed bed. They began with an old Swedish drill that happened to be on hand; it was not easy to use and so they imported an American seed drill known as 'The Farmer's Favourite'. Correll and his neighbours found that as little as 125 kg/ha of superphosphate was sufficient to increase yields to a level at which the cost/benefit ratio was adequate. Their yields were 1540 kg/ha and the cost was $AUS 1.25/ha. This was one fifth the cost of the application recommended by Custance and Lowrie. Correll and his fellow farmers also noted the savings that came from placing the seed and fertiliser together in the seed bed with the seed drill by comparing the results with those obtained when they used a spinner. They found that whereas the spinner covered 16 ha per day against the seed drill's 6.5 ha per day, the economies in seeding rate and amount of fertiliser made possible by the efficiency of the drill more than made up for the extra time taken. All the seed and fertiliser were utilised when drilled whereas the wind tended to disperse and waste both seed and fertiliser when it was spun out. The seed drill cost $AUS 70 but farmers considered that it was well worth it (Correll, 1896, p. 7). In 1898 over 20 000 ha were sown to wheat plus superphosphate using seed drills. In 1900 farmers bought $AUS 70 000 worth of super, and by 1910 farmers were applying superphosphate to their crops at rates of between 78–100 kg/ha on 80% of all land cropped throughout the state (Williams, 1974, p. 285).

Today approximately 90% of cereal and livestock producers use superphosphate when seeding cereals and apply it directly also onto pasture. In 1898 (two years after Correll and his neighbours had carried out their demonstrations) some rather inconclusive trials were carried out at Roseworthy comparing yields from seed that had been sown with a drill and seed that had been broadcast. Superphosphate was applied at a rate of 250 kg/ha with an additional 67 kg/ha of sulphate of ammonia. The drilled seed gave a slightly higher yield than that broadcast, but there was a decreased yield from the trial in which seed was drilled and the fertiliser broadcast (Lowrie, 1898, p. 479).

The campaign for bare fallow and deep ploughing

While superphosphate increased crop yields, it did nothing to increase soil nitrogen or assist moisture retention. A cultivated bare fallow was proving effective in the Northern Hemisphere and a program to explain it and persuade farmers to adopt it was begun.

By the end of the nineteenth century, farmers in South Australia had already evolved a unique form of mechanised dryland farming before the new scientific technology of Northern Europe and North America was introduced to them. They had something to measure against well meant technical advice. They were, on the whole, sceptical about advice based on farming systems developed for conditions so radically different to theirs because their empirical experience had taught them that it was wise to be so. However, many South Australian farmers who used weedy or rest fallow between cereal crops became interested in the use of a bare fallow when its presentation as a dust mulch to conserve moisture – an idea conceived in the dry farming regions of America – caused them to believe that it had in fact been developed in response to conditions similar to those they faced.

The 'selling' of bare fallow

The success of the fervent program by Roseworthy College and the Department of Agriculture to persuade farmers in all districts that it was essential to have a year of bare fallow before a wheat crop was for many years considered a great victory by the scientific community. Yet the question of why they championed the suitability of a bare fallow to large parts of the South Australian dryland farming wheatbelt with its soils that were subject to erosion, its past 'wheatsickness' due to lack of organic matter, and the unpredictable and too often low rainfall, particularly in the marginal zone, begs an answer.

The second Principal of Roseworthy College (Lowrie), continued Custance's campaign to persuade farmers to plough deep and led the campaign to establish bare fallow before wheat. Remember that Lowrie and his colleagues and successors were employed to undertake rigorous scientific investigation and to apply rational thought to farming and were expected to supply an intellectual element to partner the practical innovations of the farming community. Yet the reasons for the introduction of bare fallow, the trials carried out at Roseworthy and elsewhere to support its place in the farming system, and the subsequent insistence of scientists and technical advisers that it be used long after it had proven to be destructive, reveals an absence of scientific exploration and rational thought on the part of the 'experts' that is astounding. The task given the scientific establishment was to find some way of introducing organic matter and fertiliser (which was broadly thought to be necessary) to farms so that erosion would be lessened and yields increased. By 1910 farmers were using superphosphate to provide a boost to fertility and getting good wheat yields as a result. Many of them were also exploiting the medic that volunteered in their rest fallow by establishing sheep flocks on their farms. This was an enterprise that was compatible with growing cereals and it gave them another source of income. The extended grazing provided the animal manure so talked about in the early days as a means of fertilising the soil. Why then did the technical experts not pick up the obvious benefits from this emerging system and support it rather than making determined efforts to have farmers replace the weedy, or rest, fallow with a cultivated bare fallow that would keep the ground clear of vegetation for a whole year before the next wheat was sown?

'Rest' fallow

The 'rest' fallow that many farmers used left a previously cropped paddock to grow natural pasture while they planted their cereal crop on another paddock. By 1880 the size of wheat farms averaged 260 ha and in the more marginal areas they were as large as 400 ha, so it was possible to rest land. Before farmers began using superphosphate the benefit to the subsequent crop from the rest fallow was not particularly significant. More significant was the fact that the growth of natural pasture (relatively poor though it was, particularly on marginal land) at least provided ground cover and some defence against erosion and, when the land was eventually cultivated, there was a little nitrification from the organic matter. Of course it was not enough to supply the fertility needed for consistently high yields of cereals and a gradual decline in fertility took place and by 1880 had become a serious problem.

Nonetheless, it was the continued use of rest fallow even by many farmers who ploughed a bare fallow before a wheat crop (a three year rotation rather than two) that probably saved the state from a total collapse of its arable land. In 1900 a farmer, W. Lehmann, reported to his branch of the Agricultural Bureau that he had observed the extremely good growth of naturally occurring 'clovers' on his farm after using super on his wheat and leaving the land in a rest fallow the following year. He had invested in sheep to make use of this good pasture and suggested that those farmers who were experiencing the same good growth should buy and run some sheep (JAI, 1900, pp. 732–3). Many farmers quietly took his advice.

Some evidence for the widespread use of rest fallow can be found by examining farmers' responses to a cropping survey carried out by a South Australian Royal Commission in 1868–9. The intention was to assess the degree of fallow carried out by farmers. Many of those surveyed volunteered detailed responses rather than just the straight yes or no asked for, and these answers revealed that while most farmers used fallow they did not run a strict wheat/fallow/wheat rotation, and neither did they use a bare fallow. The ambiguity lay in the definition of the word 'fallow'. Most farmers were, in fact, letting weeds grow (a weedy or 'rest' fallow) to rest their land while they grew their cereals on another part of the farm in order to try to return some fertility to the soil. A farmer named Sanders responded that while he used no manure he was constrained 'not to overtill but to fallow every alternate year'. (In referring to this tendency by farmers to enlarge on simplistic requests for information in order to reveal reality, the economist Edgar Dunsdorfs later made the point that 'greater realism than that of official statisticians was shown by practical farmers who approached statistics with commonsense' (Dunsdorfs, 1956, pp. 138–43).

English and American bare fallow

Following the introduction of the steel mouldboard plough in England, farmers were able to plough up to 30 cm deep and more and turn weeds into the soil. The deep ploughing opened the heavy wet soil to the air and the ploughing in of the organic matter released nitrates into the soil. This fallow was ploughed in the autumn following the cereal harvest and it left the land cultivated and unused throughout the winter. The seed bed for the subsequent wheat crop was prepared in the following spring. This fallow did increase yields through the more efficient suppression of the heavy growth of weeds that is common in Northern Europe and the nitrification of the soil from the ploughed in organic matter. Yields were further aided by the addition of nitrogen and phosphate fertiliser. Initially, it was this ploughed fallow that Custance and his successor Lowrie urged the South Australian farmers to adopt, but it depended on deep ploughing and the farmers were not interested. There seemed little point in adopting a practice that was essentially an operation carried out to deal with heavy moist soil when the soils on which South Australian farmers grew wheat were dry for most of the year. Once again the difference in climatic conditions appeared to be unheeded by the technical experts. The ploughed fallow was developed for a winter fallow in preparation for a crop that would be seeded to grow in late spring and summer; in Australia the fallow period was over summer in preparation for a crop that would be seeded to grow throughout the winter.

When the bluff and friendly Lowrie succeeded Custance as Principal of Roseworthy, he established a series of trials to illustrate the benefits of deep ploughing and bare fallow before a wheat crop and these convinced him that South Australian farmers should adopt it to reverse their declining yields. He included heavy dressings of superphosphate in his trials (JAI, 1897a, pp. 17, 711).

Lowrie distinguished between two types of fallow in his subsequent 1897 message to farmers. He advised those in the relatively high rainfall zone of between 400 and 450 mm to use the 'English' ploughed fallow through the winter and grow spring and summer crops of 'mangolds, kale, rape, white mustard, maize, sorghum and millet'. He advised farmers in the drier zone where rainfall was below 400 mm to plough an American 'bare' fallow which began in the late winter or spring and carried over throughout the summer until sowing began in the following autumn (JAI, 1897b, p. 137). Farmers yawned and took no notice because they were being asked to undertake an intensive farming program that was beyond the resources of most of them. They had tried these methods and they had proved impractical on their farms.

Moisture retention due to bare fallow

In 1902, however, a new factor was introduced into the list of reasons being put forward to persuade farmers to adopt bare fallow. Dr Lowrie was succeeded as Principal of Roseworthy by an American, Professor J.D. Towar of Michigan. He was enthusiastic about the idea that his compatriot Hardy W. Campbell had developed in 1892 of cultivating land in dryland farming conditions so that a dust mulch was created on top of the soil in order to conserve moisture from one winter to the next. The theory was that by breaking up the surface of the soil, the capillaries that brought the moisture up from below would be destroyed and therefore the moisture from one winter's rain would remain under the surface ready to be taken up by the plant when the next winter's crop was planted (Williams, 1974, pp. 292–3; JAI, 1902, pp. 335–6, 521–2).

The promise of moisture conservation in dryland conditions was too good to resist and farmers began to follow the band.

Thus there came forth from Roseworthy a curious hybrid message. It contained some of the components of the English ploughed fallow – plough deep, open the soil (but not for evaporation as it was in England, somehow on the journey to Australia the reason had become for water penetration), and aerate the soil to promote biological activity. To this was added another attraction. This was the inducement of moisture conservation through the elimination of weeds and the constant tilling of the dust mulch, yet there was no proof that moisture would be conserved in the relatively poor soil of the low rainfall districts below 400 mm where the majority of South Australia's wheat crops were grown.

It is difficult to tell whether the use of bare fallow and a dust mulch did leave behind enough moisture from winter rains to assist in the crop a year later.

We have found no records of moisture retention trials being carried out at Roseworthy during the Lowrie, Towar regime to establish that moisture conserved under a bare fallow/dust mulch in the South Australian wheat belt led to a quantifiable yield increase. Custance, Lowrie and Towar seem to have been promoting their own opinion based on an opinion originating in the United States.

In the next decade when work was done in the United States to test the theory ' . . . the results of investigations in the United States, summarized in 1915, showed that the dust mulch was not a factor in moisture conservation under normal wheatbelt conditions' (Callaghan & Millington, 1956, p. 116).

The costs of bare fallow

In 1898 a South Australian farmer named Coulter recorded that the cost of any possible moisture retention due to bare fallow was too high for farmers. Bare fallow, he pointed out, required two additional cultivations of the soil and a loss of seven months grazing and there was no proof of any significant cereal yield increase to compensate (Coulter, 1898, p. 379).

In 1945 another farmer, J.A. Kelly, reported to his fellow farmers that the claim that moisture was conserved under fallow was a fallacy. There must have been some trials going on at Roseworthy at this time because he referred to trials there that substantiated his claim. It was his opinion that it was wasteful leaving the land bare and unproductive for a whole year because now that most farmers used tractors it was not necessary to take a whole year to carry out a tillage program for the cereal crop. He added that bare fallow, far from conserving moisture for the subsequent crop, encouraged soil drift, water erosion, loss of organic matter and oxidisation due to the frequent cultivations involved. The loss of the total plant matter meant that no organic matter or useful bacteria remained in the soil (JDA, 1945, pp. 455–9).

Callaghan and Millington wrote in 1956 that 'fallowing was introduced and used on the false premises of conserving moisture by the formation of a well tilled surface mulch ... ' (Callaghan & Millington, 1956).

In 1963 the Department of Agriculture carried out a review of the moisture conservation potential of bare fallow and concluded that in lighter sandy soils the gain was a negligible 2–3%, although on high rainfall, self-mulching soils it could be up to 20%. However, this retention depended on a particular concatenation of circumstances to come about – for example, good rainfall late in winter, and an absence of hot dry winds in summer, but most of all a relatively high rainfall zone which was a rare occurrence in the South Australian wheatbelt (French, 1963, pp. 42–8).

In 1987 Professor A. Dahmane told an FAO Expert consultancy that work on soils in Tunisia similar to those in South Australia had led to the conclusion that the small amount of moisture conservation that occurred during the fallow had dissipated by the time the succeeding crop was planted, thus nullifying any benefit (Dahmane, 1987, pp. 7–9).

Weed and disease control

The control of weeds during the bare fallow phase on farms in the Northern Hemisphere certainly helped increase cereal yields. But in South Australia, certainly in the early years, weed control was not a great problem. The naturally low fertility of the soil prior to the widespread adoption of superphosphate after about 1910 did not provide a basis for heavy weed growth. Most of the common weeds in crops in South Australia have been introduced over the past eighty years or so, and one finds little discussion among the early farmers about weed problems in crops. Mechanised shallow cultivation and grazing during the pasture phase provided an effective way of controlling weeds on dryland farms. Rather than improving the soil structure, the rigorous destruction of the weeds and volunteer pasture that bare fallow and a monoculture of cereals accomplished led to a reduction of organic matter available to the soil over time and an eventual decline in the quality of the soil structure thus both reducing its ability to absorb rainfall and its fertility.

There was some belief that bare fallow had an effect on the control of 'takeall', a disease of wheat, and it may well have been effective in this matter (JAI, 1901, p. 62).

By the time it became more widely known that eliminating grasses (particularly barley grass) was effective in breaking the cycle of root disease in cereal crops, herbicides were available to selectively destroy grasses without harming the medic or sub-clover pasture. The rotation with medic and sub-clover pasture with cereals, and the grazing that takes place during the pasture phase, keeps grasses to a minimum and is also a recognised means of breaking cereal disease cycles.

Nitrification of the soil

Farmers also found attractive the claims for bare fallow that it could increase the nitrification of the soil. The quest of the South Australian dryland farmer was for some form of organic matter to enhance the quality of the soil on the farm. The whole thrust of both the dust mulch and bare fallow was away from this. The effectiveness of the dust mulch proposed by Professor Towar depended on a total denudation of the soil through the late winter, spring and summer so that the absence of plant material would prevent the drawing up of moisture from the sub-strata. In order to sustain the dust mulch farmers had to plough early and then constantly cultivate or harrow out any volunteer plants that germinated if unseasonal rain occurred. This meant that not even accidental organic material was available to help sustain the soil structure or provide a modicum of fertility beyond that which the application of superphosphate could provide. Over a number of years the constant use of bare fallow reduced the seed available for volunteer pasture. The germination of pasture declined and the production of organic matter began a downward spiral that intensified with each season of bare fallow. By destroying what organic matter the soil could produce the bare fallow made the soil even more vulnerable to wind and water erosion.

Barley without bare fallow

While claims have been made that the increases in wheat yields that were recorded between 1900 and 1930 were due to the effectiveness of bare fallow in contributing nitrification, weed control and moisture retention, it is interesting to contrast the experience of those farmers who sowed barley on their farms. Many farmers, particularly on Yorke Peninsula, never followed the advised 'short' rotation of bare fallow/wheat. Barley for them was always an important cereal crop and many used a rotation that ran bare fallow, wheat, rest fallow (during which natural pasture covered the paddock and was grazed by sheep), and barley, then often another rest fallow, then bare fallow and wheat again. Barley was grown to supply the demand for malt barley that resulted from the local demand for beer and also for feed grain for pigs and poultry. The custom of growing barley for grazing was not common.

In considering this long rotation a curious fact emerged during our research into the story of bare fallow. After diligently searching many sources of farmer practice and technical advice it suddenly struck us that there did not seem to be any concern that barley followed rest fallow, yet for over fifty years the advice from scientist and technical adviser was that wheat should never be grown without a preceding bare fallow. It was always bare fallow before wheat but not before barley. It may be that the practice of sometimes sowing barley after wheat grown on bare fallow was believed to carry the benefit of the bare fallow over to the second crop.

Barley yields after the introduction of superphosphate were never described as being in decline due to 'barley sick' land, and there never seems to have been any concern that the land after the rest fallow was so weedy that it had a deleterious effect either on the moisture available or on the subsequent barley yield.

Was bare fallow effective?

While the rationale offered by the experts for adopting bare fallow in South Australia appears to be fallacious, initially the introduction of bare fallow before wheat did seem to be producing better yields. Yet the adoption by many farmers of bare fallow at the same time that they began applying superphosphate to their crops leads to some ambiguity about the cause of the increases. Certainly bare fallow before wheat produced comparably better yields than wheat after wheat in trials (Spafford, 1918). The farmers' difficulty in distinguishing what yield increase was due to bare fallow and what to superphosphate was compounded by the fact that no trials were made of wheat immediately after pasture or rest fallow. The standard experiment conducted by the Department of Agriculture compared bare fallow/wheat with wheat/wheat or with bare fallow/wheat/barley or oats and rest fallow/bare fallow/wheat (Spafford, 1918, pp. 11–24).

In spite of any real proof, the belief of the technical establishment in the efficacy of bare fallow, reduced to the simple slogan 'bare fallow before wheat', became the mantra which was recited to persuade farmers to adopt the practice of bare fallow in order to reach the nirvana of higher wheat yields.

The costs of bare fallow

At the same time that the scientific and technical community seized upon bare fallow as the answer to wheatsickness and were blazoning the formula of bare fallow before wheat across the state, there were farmers who were counting the cost of the bare fallow and using a different source of fertility for their crops and they were being reported in the Journal of Agriculture. In the opinion of some farmers, the time taken to work the bare fallow had to be weighed against their commonly used seed bed preparation program. In 1898, a farmer named Coulter reported that the adoption of bare fallow meant that farmers had to work the land in spring to clear the land for summer and then again in autumn to sow the subsequent crop. He provided figures to support his case that the extra work load cost more than the potential advantages of a ploughed fallow even when the cultivation was done with a relatively shallow plough. It took twelve weeks to plough 121 ha with a three furrow plough for a fallow, but only five weeks to cultivate the same area with a light scarifier and a further two weeks to harrow it before sowing the cereal seed. He himself had five paddocks of 60 ha each. When experimenting he had sown three paddocks to cereals, put one to bare fallow and left one in pasture (rest fallow) (Coulter, 1898).

Other farmers reported another disadvantage if they used a long bare fallow. In 1898, Joseph Correll noticed the good growth of the 'native clovers' after dressings of super on wheat crops (JAI, 1898, p. 208). The grazing of sheep in the rest fallow reported by many farmers led them to ask if it made sense to plough this up.

The consequences of bare fallow

It has been claimed that by 1909 well over half the farmers in the state had made bare fallow before wheat part of their normal practice (Williams,

1974, Table XV, p. 291). It is difficult to determine how many of those questioned were actually using the bare fallow alone and how many, when responding to the broad question 'Do you use fallow?', answered without distinguishing between bare fallow before wheat and the rest fallow that was part of a longer rotation. The same scepticism that Edgar Dunsdorfs expressed towards an earlier survey to determine land usage on cereal farms can be applied here, and research suggests that such scepticism may well be justified.

In 1911 A.J. Perkins (Deputy Principal of Roseworthy College) when reporting the field experiments of Roseworthy farm mentioned that the 'common enough rotation north of Adelaide (by neighbouring farmers) was bare fallow/wheat followed by pasture'. He quoted figures showing that although this rotation was not as immediately profitable as the short rotation of bare fallow/wheat favoured by the Department of Agriculture, the farmers' retention of a pasture phase ensured that because there was a return of humus to the soil after the cropping phase it produced better fertility and thus increased profit over time. He noted that most farmers, particularly in the Lower North district (around Roseworthy) were carrying three to four times more sheep as a result of this rest fallow or pasture phase and he believed that 'these facts are now so generally admitted . . . as to stand in no further need of experimental demonstration' (Perkins, 1911, pp. 11–12).

In 1918 he told the Kybybolite Agricultural Bureau that in spite of the fact that Roseworthy and the Department of Agriculture were still actively promoting bare fallow, he himself believed that 'Signs are not wanting that this rotation has had its day' (Perkins, 1918, pp. 763–75). He advised farmers to adopt longer rotations and forage crops. Yet in 1929, Dr A.E.V. Richardson, the Director of the recently established Waite Institute (the agricultural faculty of the University of Adelaide), told farmers that bare fallow was necessary and that a 'thorough working of the fallow' was best (Richardson, 1929, pp. 297–317).

By 1934 Perkins had also become a champion of bare fallow and he wrote that 'There is no other practice that has had a greater influence for good on our mean yield per acre . . . fertilisers etc. count for little or nothing unless the land has been fallowed in the previous season'. He even called for the return of the mouldboard plough and told farmers that although they may well have reasons for shallow cultivation they should try to achieve a gradual increase in the depth of cultivation (Perkins, 1934, pp. 696–717).

Although Perkins had changed his mind about bare fallow, it was clear by 1935 that there was some concern within the technical community that their continuing advice that bare fallow was essential for good yields of wheat was in conflict with the evidence they were seeing on farms of its destructiveness when the advice had been wholeheartedly followed. The young R.I. Herriot was among those who provided evidence that suggested that the reasons given for years for the use of bare fallow were no longer viable. He reported trials at the Waite Institute that showed that fallowing for moisture did not work, and that the nitrogen for the crop could come more reliably from the medic or sub-clover that grew in the farmer's 'rest' fallow phase (Herriot, 1935, pp. 283–4).

In 1937, forty years after bare fallow had been incorporated into dryland farming in South Australia, a Parliamentary enquiry found serious soil erosion in the heavier soils of the Upper Northern areas and the mid-North and in the light soils of the Murray Mallee where the adoption of bare fallow before wheat had been greatest. Little soil erosion was apparent in the lower half of Yorke Peninsula where farmers grew medic pastures in rotation with barley (Williams, 1974, p. 308). In the Hundred of Belalie in the Upper North, for instance, it was reckoned 76% of the arable land had lost more than a quarter of its surface soil and gullying was severe (Williams, 1974, p. 310). The report of the Parliamentary enquiry unambiguously stated that 'the practice of prolonged and continuous fallowing was the culprit' (Williams, 1974, p. 305). In spite of the results from the trials at the Waite Institute, the conclusions of the Parliamentary enquiry, and the growing doubts about the benefits of bare fallow among the more observant technical personnel, A.R. Callaghan in 1938 reported that the Roseworthy College farm was still growing all its wheat on bare fallow (Callaghan, 1938, pp. 558–73).

By the early 1940s, the adoption of bare fallow before wheat was both at its peak and beginning its decline. By 1940, 93.7% of farmers in the Lower North district were recorded as using bare fallow before a wheat crop and in all districts, except the high rainfall South East where little wheat was grown, the percentage was high. At the same time it was recorded that by 1940 many farmers were interspersing a year of volunteer pasture (containing a large proportion of medic) into their rotations and some were deliberately sowing medic and clover in order to benefit not only from the improved grazing it provided for their sheep flocks but also for the fertility it provided for their cereal crops (Bowden, 1940, pp. 399–416; Breakwell & Jones, 1946, pp. 252–6).

The trouble was that in spite of evidence on all sides of the destructive effect of bare fallow and the constructive effect of the farmers' system, the technical hierarchy of Roseworthy and the Department of Agriculture simply could not bring itself to admit that its advice had been misguided.

The obsession with bare fallow among the scientific and technical community remained, and one sees the result in confused messages that began to emerge in extension material circulated to farmers.

In 1943, R.I. Herriot wrote in the *Journal of Agriculture* that a survey in the County of Victoria (the mid-North of South Australia) showed that on wheat land where a bare fallow/wheat rotation had been followed only 24% of land had more than 75% of its soil intact. Herriot listed the consequences of the continual use of bare fallow before wheat – the eventual result of loss of soil structure, poor water penetration, a cloddy seed bed and erosion. It now seemed that the result of bare fallow was directly the opposite to that claimed for it at the beginning of the campaign. Herriot, convinced by what he was seeing on farmers' paddocks, called for more pasture phases to be included in rotations. Yet, his concluding remarks were that 'Fallowing is essential for wheat production in most districts' (Herriot, 1943, p. 55).

In 1946, E.J. Breakwell of Roseworthy College reported that farmers were abandoning bare fallow and growing medics and sub-clovers to replace the fallow phase before the wheat. He personally believed that this farmers' system was suitable only for the high rainfall areas, but he admitted that there were indications that it may succeed in relatively dry areas. He believed there was 'too much emphasis' being placed on the role of bare fallow before wheat. In spite of the fact that it was his colleagues who determinedly carried on the campaign for the use of bare fallow, he blamed farmers for not being more progressive in this matter. Clovers and medics are 'the life blood of the cereal areas' he wrote, yet 'farmers persist in ploughing them under before they have time to perform nature's allotted task – that of adding to soil nitrogen' (Breakwell, 1946, pp. 188–90).

One could have hoped that scientists asked to bring 'rational thinking' to South Australian dryland farming would have led the way in exploring the possibilities inherent in the farmer reports of the benefits of legumes appearing in their rest fallow, but they did not. They continued to hanker after deep ploughing and fallow although it had never become a serious alternative for farmers in the state.

It was not until 1954 that R.I. Herriot admitted on behalf of the scientific and technical establishment that after seventy years of trying 'We have not been able to show that deep ploughing is necessary or economic for cereal growing, nor is there anything to be gained from nicely inverting the furrow slice' (Herriot, 1954, p. 91). He was quoted later as saying 'The aim should be to reach the goal of ideal tilth by seeding time with the smallest number of workings by the least destructive machines' (Callaghan & Millington,

1956, p. 123). This was why in the last decade of the nineteenth century farmers had refused to abandon rapid, shallow tillage in favour of deep ploughing favoured by Custance and Lowrie among others. Even when bare fallow had become discredited the technician's belief in its effectiveness as 'good farming' continued.

Callaghan and Millington wrote in 1956 that '. . . Unfortunately fallowing, and especially the excessive tillage that has gone with it, has caused serious deterioration in soil structure', yet both authors still supported bare fallow because they claimed it increased the nitrification of the soil (Callaghan & Millington, 1956, p. 120).

The truth of the matter was that the nitrogen in the soil required for the crop was available from the abundant legume pasture that grew before the crop, and to impose bare fallow on this was both costly and unnecessary.

The rotation on the research farms

The research farms managed by Roseworthy College and the Department of Agriculture continued to plough a bare fallow before wheat. Minippa Research Farm was the first to break the mould and, in 1953, wheat was grown after pasture instead of after bare fallow, (Day, 1954, p. 162). Until then the medic at Minippa was cut for hay thus destroying the seed supply for the next pasture and it had to be re-sown each year. The seed for the new pasture was grown and harvested on the farm. On the other side of the research farm fence, neighbouring farmers put their sheep to graze on their medic paddocks and managed that grazing so that the seed reserve was safeguarded and natural regeneration of the pasture (which cost the farmer nothing) occurred. Farmers had to be more economical in their use of labour. The medic supplied most of their needs and they rarely made hay. In spite of the change at Minippa, R.N. McCulloch reported in 1956 that all the wheat being grown on the Roseworthy Farm was sown after bare fallow (McCulloch, 1956, pp. 267–72).

All wheat grown on the research farm at Turretfield was grown after bare fallow until in 1958 when a 'clover ley' replaced bare fallow (JDA, 1958, pp. 208–9). This was twelve years after Breakwell reported farmers using this rotation and fiftysix years after Lehmann and other farmers reported their own discovery of its value.

There was no need for bare fallow. Improved soil structure and soil nitrogen came from the legume pasture phase, not the bare fallow. The scientific contribution to the debate about the use of bare fallow on dryland farms was not tremendously helpful. The Waite Institute was established in

Table 3.1 *Mean yield of wheat in various rotations at Waite Institute, Adelaide, 1926–83*

	Mean yield of wheat (kg/ha)	
Rotation	1926–51	1952–83
Continuous wheat	874	692
Fallow–wheat	2300	1403
Peas–wheat	1668	1421
Oats–fallow–wheat	2340	1829
Pasture–fallow–wheat	2663	2063
Oats–pasture–fallow–wheat	2999	2536
Barley–peas–wheat		1916
Pasture–pasture–wheat		2033
Pasture–pasture–fallow–wheat		2402
Four pasture wheat–wheat		
wheat 1st year		1822
wheat 2nd year		1473

Source: Norton & Britza, 1982–3, pp. 221–7.

1924 under the terms of the will of Peter Waite, a grazier who gave his own mansion and land in the fertile foothills of Adelaide to the University as well as a generous endowment in order that the Institute be used to investigate the natural pastures and their potential. Too soon the Waite became a tool of those who considered the selection and breeding of cereals to be the most important priority for the state's agriculture and so it has remained until today. A long term rotation trial (see Table 3.1) to show the differences in yields that could be obtained from various rotations was begun at the Waite Institute in 1925 on half acre plots under an average rainfall substantially higher (600 mm) than that of the cereal zone (200–400 mm).

It was not until 1947 that several of the earlier rotations were discontinued to make way for rotations that included legume pasture, although a pea–wheat rotation (much favoured by Roseworthy) was introduced earlier. By 1983 the research team concluded that

The inclusion of a pasture phase in the rotation, whether oats for grazing, or Wimmera Rye Grass and subterranean clover, has generally resulted in higher yields of wheat than those obtained from rotations without any pasture. However, the yields show considerable variation particularly in the rotations without fallow following the pasture phase.

It should be noted that until 1973 a mouldboard plough was used in these trials to work the fallow and that only after that time a 'standard spring release cultivator' (scarifier) was introduced 'to prevent erosion'. The trials show the gradual loss of yield over a long period when a fallow is in constant use, namely, fallow–wheat–oats–fallow–wheat and pasture–fallow–wheat.

The trials also show that the intervention of a fallow into a pasture-based rotation can provide some benefit (a yield increase of 369 kg in the case of pasture–pasture–fallow–wheat compared to pasture–pasture–wheat) but this means the loss of a year's sheep production and extra costs of labour and energy. In fact, the old argument put forward by the farmer raises its head. Is the cost of the contribution of the fallow in the form of a relatively small increase in the yield of cereals more than it is worth?

The major flaw of the experiment is that it reflects the practice of Roseworthy College and the other research farms and not that of farmers who were getting good results from wheat after medic pasture in the 1930s. It was not until after 1947 that the Waite Institute, endowed to investigate the improvement of natural pasture, incorporated additional rotations that included legume pastures. It is doubtful if the pastures used in the trials were the medics that are used in the rotation being practised by most farmers in the cereal zone. The composition of pastures (which are indeed the appropriate ones for the climatic conditions at the Waite which is in the foothills of Adelaide) shown in the explanation of the trials refer only to subterranean clover, Wimmera Rye grass, lucerne, and phalaris.

Medic and wheat yields

One could say that the medic/cereal rotation was the result of a series of chance happenings – the accidental introduction of the plant, the adoption of shallow cultivation because of shallow stony soil and difficulty in clearing land, the superior performance of superphosphate *vis-à-vis* nitrogen in providing an effective fertiliser for cereals in dryland conditions, and the opportunistic exploitation of the medic by farmers with little cash and less labour.

Once the medic/cereal rotation was used on most dryland farms, average yields of wheat on farms reached 1.5 tonnes/ha and rarely went below. In 1982 Professor Donald of the Waite Institute drew a graph which he claimed showed that productivity increases on cereal farms since 1940 had been made possible by the adoption of annual legume pastures (Donald, 1982, fig. 3.1, p. 63).

He also calculated that the nitrogen contribution to the soil between 1971 and 1976 from permanent legume pastures was equivalent to 1.5 million tonnes of nitrogen and another 0.5 million tonnes from legume pastures grown in rotation with cereals (Donald, 1982, pp. 77–8). The government statistician paid tribute to the system by claiming in 1985 when the principal wheat districts with rainfall between 200 and 350 mm had an average yield of 1.66 tonnes/ha it was due to 'the adoption of nitrogen building legumes in the rotation' (SAYB, 1985, pp. 370–1).

A farmer, K. Beare, tells of his family farm of 600 ha in the wheatbelt, where average rainfall is 375–400 mm. In the 1920s his father used a wheat/bare fallow rotation. In the 1930s falling yields led him to incorporate a rest fallow during which medics volunteered. By the 1940s he was sowing 'barrel' medic and applying superphosphate to the pasture and began to abandon fallow. In the next ten years the productivity of the farm more than doubled. In the three years to 1976 production from 600 ha averaged 6000 kg of wool, 6000 kg of meat, and 420 tonnes of grain (Webber, Cocks & Jefferies, 1976, pp. 30–1).

Why did so many farmers adopt bare fallow?

In view of the obvious benefits to dryland farmers of a legume pasture phase, why did so many farmers adopt the bare fallow being advised by the technical experts? In the first decade of 'scientific agriculture' in South Australia, farmers were sceptical and resentful of the technical and scientific expert. When Correll succeeded in finding a means of using superphosphate on dryland farms that was compatible with the resources of most farmers, one would have thought that it would have intensified this scorn of the theorist. If one reads the *Journal of Agriculture* and the subsequent histories of farming one is given the impression that the higher yields that followed the adoption of superphosphate and the initial adoption of bare fallow before wheat reversed the position of the technical expert *vis-à-vis* the farmer and made the technical expert very influential. By 1910 many farmers appeared to take seriously the advice coming from Roseworthy College. For the next thirty years they vigorously debated shallow cultivation versus deep ploughing at farmers meetings and in the pages of the *Journal of the Department of Agriculture*, even though few of them ploughed deeply for more than a year or two. They discussed the merits of bare fallow – some solemnly assuring their fellow farmers that they had 'saved' up to 3 or 4 inches (75–100 mm) of moisture from one season to another – something even the technical experts must have found an exaggeration. Farmers had tried to grow kale, mustard, beans, rape and turnips, and they tried again but they failed to make such crops profitable.

They tried to grow lucerne between wheat crops but the bare fallow destroyed it. Over time, Roseworthy and the Department of Agriculture became more interventionist. They became the major source of extension advice, regulation and administration and they set up research farms at Booborowie, Kybybolite, Minippa, Turretfield, Struan and Kingsford and of course the farm at Roseworthy College. On these farms they grew their wheat after bare fallow, turned their oats and barley and lucerne into silage, fed turnips and kale to sheep and made hay from the native pasture that volunteered while the neighbouring farm sheep simply grazed it in the field. They continued to try to grow English grasses for hay without much success, and they continued to try to persuade farmers to improve their natural pastures by oversowing with these exotics. The rapport that developed during the campaign of superphosphate and bare fallow did not last long, but the status of the technical expert became enshrined in the mystique of scientific terminology where it remains even today. This mystique is diluted in Australia by the development of strong, independent farmer groups such as the Australian Farm Management Society and the Kondinin Farmers who provide an alternative source of exploration and advice to their colleagues.

Why did the scientific community persist with bare fallow?

It seems strangely obtuse that technical experts and scientists should have chosen so dramatically to persist with bare fallow and conserved fodder as the basis for dryland farming when a simple, cheap solution to farm productivity was being demonstrated by farmers all around them. The masking of the destructive result of bare fallow by the fertility induced by superphosphate may explain some of this obtuseness. Another reason may be the carefully constructed wheat yield competitions that for many years gave the prize to crops sown on bare fallow. Eventually (by 1937) even the superphosphate could not compensate for the loss of soil structure and soil fertility and so the huge drifts of sand and the wounds and scars of erosion appeared that caused widespread alarm in the wheatlands where bare fallow was used and no rest fallow intervened. Amazingly it seems that those scientists and technical experts who led the climate of opinion about such things simply did not understand the dangers of bare fallow in a dry climate until the destruction was visible for all to see. The first fifty years of farming in the state had shown the South Australian farmers that the harsh climate and fragility of the soil required radically different farm operations from those that existed in Northern Europe, and the way in which they gradually felt their way into a medic/cereal rotation was a further response to this experience. Yet their innovations and intuitions seem to have been ignored by those who introduced and supported the adoption of bare fallow. It is not that these early technical and scientific experts were wrong that is the cause for criticism, it is that they never seriously questioned whether they were right or wrong, and they continued to advise farmers to follow their assertions long after farmers and their own research showed that they were wrong.

It is not as though the exploitation of the medics and sub-clovers took place in secret. Farmers made their individual experience publicly available and argued the pros and cons of 'their system' against that being advised by technical experts, but such was the reversal of roles of farmer and technical adviser from about 1910 that the opinions and experiments of farmers were easily ignored. However, the technical experts did begin to abandon bare fallow as part of the research program in 1953 and by the end of the 1950s had accepted that the farmers' use of medic and clover pasture as the basis for soil fertility for cereals and year round grazing for sheep was a sound one. It seems that it was always difficult for those with research farms as their base to understand the constraints of cost and labour under which the average farmer operated. The results obtained from trial plots, which so often were the only data available to support the advice of the technical expert, were, by nature of their education, more convincing than what took place on the neighbouring farm. The technical experts in Australia had great faith in the new techniques discovered by the scientific community in Northern Europe. Yields and livestock production had been vastly improved there as a result of farmers adopting these and it simply did not seem relevant to their own search for answers to farming problems to consider seriously what farmers over the fence in South Australia were finding out. On the other hand, the data from a trial plot were so far from the every day experience of the farmer that the example of a neighbouring farmer's success had more effect than an address from a technical expert or a paper published in the *Journal of Agriculture*.

The Department of Agriculture after 1960

Once a medic/cereal rotation was adopted on the research farms, scientists and technical experts produced an avalanche of papers and addresses describing the whys and hows of the medic rotation. The degree of nitrogen made available to the crop from the medic was measured, the stocking rate that could be maintained on legume pastures in various districts was tested, diseases and insects were given special attention (notably sitona weevil that unfortunately still defies a totally successful solution) and the search for new varieties and cultivars of medics and sub-clovers was intensified. This resulted in several new cultivars that extended the use of medic and sub-clover pastures into drier areas than had originally been thought possible.

Farmers and scientists cooperated successfully when, in 1979, the blue-green and spotted-alfalfa aphid (believed to be introduced from North America) caused severe destruction initially of lucerne (alfalfa) pastures and then of medic pastures. Eradication campaigns based on frequent spraying of insecticide were too expensive for most farmers and so Dr E.D. Higgs, a pasture agronomist with the Department of Agriculture, developed a control program based on combination of a short term, low intensity spraying, the breeding and release of parasites and a longer term program for the selection of medics resistant to the aphids. Many of these latter are now available commercially to farmers. The whole affair was an outstanding response from a scientific unit within the Department of Agriculture to a crisis in the pasture industry.

One of the world's largest collections of medic cultivars was established at the Northfield Research Centre and it has become a centre of excellence, identifying and selecting cultivars for aphid resistance and for special ecological niches such as extremely arid conditions.

In the 1980s many scientists came to believe that the answer to problems with medic on dryland farms (particularly in North Africa and the Near East) lay in finding the 'right cultivar'. Dr E.D. Higgs, the pasture agronomist who headed the successful aphid control program cited above, did not agree with this. He believes such an opinion to be neither sensible nor practical. Whatever potential discoveries await the diligent plant breeder of medics

it will be quite a number of years before the first of these can possibly reach commercialisation as it takes about 8 generations to fix a medic after the first cross is made. The indicated rate of progress in breeding medics is that medic cultivars with complete tolerance to all present major problem insects are more than a decade away and in barrel and strand medic (Harbinger) this objective will not be attained for 15 to 20 years (Higgs, *c*. 1981, p. 7).

Higgs favours the immediate repairing of poor quality pasture by re-sowing a mixture of existing cultivars that allow for the differences of soil type, climatic variation and insect populations on each farm. He suggests that research funds should be directed away from the ubiquitous selection and breeding programs and used to investigate ways in which nitrogen from medics can be manipulated to benefit cereal crops. In the meantime, medic pastures are 'inexpensive to establish and maintain and are one of the world's outstanding examples of low cost biological nitrogen fixation . . . the surest path to better nitrogen balance is the path that leads to higher rates of nitrogen fixation and this means growing more medics' (Higgs, *c*. 1981, pp. 1, 21, 27).

It proved difficult to overcome the inherent tendency of the research farms to doubt the efficacy of the simple rotation. The continuing lukewarm attitude is illustrated by a report from the Minippa Research Farm in 1981. Minippa had been the first of the Department's farms to adopt the farmers' rotation in 1953, yet in 1981 the system being used on there was described in the following terms.

Minippa Research Centre has used barrel and Harbinger medics for many years, initially with great success. Cropping intensity has varied and for the immediate past decade it has not been as intensive as is currently thought necessary for good medic pastures. The medic pastures have been largely grass in recent years. The cropping rotation is now being developed towards crop and pasture in alternate years. Last year a low rate of inexpensive herbicides was applied and this year there has been a fairly dense establishment of medic dominant pasture . . . [The staff] . . . will attempt to produce some medic seed for use where needed on the farm (Higgs, *c*. 1981, pp. 20–1).

The decline of the pastures on the Minippa farm was contrasted with the enthusiasm and determination of innovative farmers in neighbouring districts who were in the same year were finding new and successful ways to renew old pasture and to exploit it in more efficient ways.

Farmers and medic after 1960

A commercial seed industry

In 1962 farmers were finding difficulty in expanding their medic and sub-clover pastures due to a shortage of commercial seed and they believed that seed merchants were profiteering from sales of sub-clover and medic seed, some of it produced on South Australian farms, but much of it imported 'sometimes even in semitrailer loads' from as far away as Western Australia (Seedco History, 1988, pp. 5–6).

A small group of seed producers decided actively to encourage other farmers to adopt seed production by forming a South Australian Seed Producers Association, and two years later when production became sufficient the South Australian Seedgrowers' Cooperative Limited (Seedco) was set up to market the pasture seed. The idea was a huge success, and by 1974 Seedco was actively seeking export markets in North Africa and the Near East and elsewhere. It was in the course of this initiative that Seedco first got involved in sending its farmer members to North Africa to teach Arab dryland farmers how to establish and exploit medic pastures on their own farms.

The Department of Agriculture's Small Seeds Section set up to support the initiatives of Seedco with research and extension as well as the regulation and certification of seed production was a reluctant gesture by the technocrats who saw it 'in the beginning (as) something of an act of faith because research results proving the idea [of pasture seed production] had not been accumulated' (D. Ragless, personal correspondence, 1989, in Chatterton papers).

Probably more useful to the success of the seed industry was the invention of a mechanical seed harvester and thresher by an innovative Western Australian farmer named Earnshaw, who built and patented a prototype that was purchased by the South Australian manufacturer Horwood Bagshaw and remained until recently the only specifically designed seed harvester. It is not an easy machine to use and most farmers make their own modifications in order to increase its efficiency.

Reinforcing and extending work on medic

A progressive group of medic farmers in the mid-North of the state in 1987/8, convinced more than ever of the value of the system and its stability in the face of harsh conditions, became concerned at the advice being given to farmers to adopt a chemical fallow and grow continual wheat, thus destroying the regenerating medic pasture. They approached both the South Australian Government and the Federal Government for funds so that they themselves could produce extension information for other farmers. They wanted to reiterate to them the long terms benefits of a medic/cereal rotation, and tell them about the improvements in management they had developed, the new markets they had found for medic pods as animal feed for live sheep shipments to the Gulf, and an innovative and more efficient means of mechanical harvesting of medic pods and of cleaning and threshing seeds that they were developing. They received no support, but, determined, they emptied their own pockets and produced a film on their own (personal communication, notes and video in Chatterton papers).

They, with Dr Higgs and others, believed that 'medics can undoubtedly

... lead to lowering of soil erosion rates, increases in soil nitrogen and organic matter, decreases in cereal diseases, increased livestock production and most importantly a prosperous and stable rural economy' (Higgs, *c.* 1981, p. 23).

Conclusion

Dryland farmers in North Africa and the Near East were persuaded by French colonialists and later by development agencies to adopt a deeply ploughed bare fallow because in Europe it killed weeds, aerated the soil and enabled nitrification of the ploughed-in organic matter to provide fertility for the subsequent cereal crop. It also destroyed the natural legume pastures that grew on farmland in the region.

Later the theory of moisture conservation as a result of deep ploughing and bare fallow was given as a reason why farmers should use this technique. Technical experts promulgated the belief that the deeply ploughed soil would encourage cereal crops to develop deep roots and spread root fibres, thus enabling the plant to find more moisture (Anon., 1987).

The farmers of North Africa and the Near East adopted deep ploughing and long bare fallow because they could not withstand the massive pressures to do so that accompanied the introduction of European and American farming technology. Their traditional farming systems could not produce sufficient grain and meat for their growing populations who increasingly demanded a better diet in which meat would play a more common part. To be modern and to produce more, they were told mechanisation was necessary. If they wanted to mechanise they had to use the deep ploughs and the accompanying implements and crawler tractors that were imported as part of the technology package which included long cultivated fallow, and nitrogen fertiliser. They were unfortunate that the machinery had been designed for the humid, heavy soils. At about the same time as mechanised deep ploughing and bare fallow were introduced, the use of nitrogen fertiliser as an additional source of soil fertility for crops was being encouraged, and these three elements became and remain the technical package for cereal production in the region. Livestock production was propped up and attempts made to expand it by using grain and conserved fodder and by introducing animal health programs. By the early 1970s the domestic demand for meat and grain was outstripping the ability of the farmers using these packages to satisfy demand and governments were forced to look for alternatives. But what alternatives were there?

In the mid-1970s the medic/cereal rotation burst onto the world stage.

The South Australian Government (spurred on by Seedco) began a program of transferring knowledge of the system to the region in return for trade. The Department of Agriculture in its first ventures into promoting the system quickly transmuted it into a 'ley farming system' and, indeed, Roseworthy and the Department had used it as a 'ley' (that is, a sown pasture cut for hay and re-sown to be rotated with cereal crops) for many years. It was an inaccurate description of the farmers' system of medic pastures and cereal production but they persist with it (Webber & Boyce, 1987).

Technical experts and scientists went off to countries in North Africa and the Near East on visits and study tours and became available as experts for projects. The Waite Institute sent Dr E.D. Carter, a pasture agronomist who specialised in medic and sub-clovers, to tour the region. He wrote a number of invaluable reports for development agencies and international research institutes in which he collated relevant information on dryland farming in both Australia and North Africa and the Near East and described a vision of a brave new farming world where medic and sub-clover pastures would replace the bare fallow of the Arab farm and bring huge increases in production (Carter, 1974, 1975, 1981). Projects with the objective of showing farmers in the Arab world how to exploit medic were not long in being established. Both expert medic farmers and technical experts were employed to work on some of these projects and once they were overseas their different methods and attitudes again emerged and the results can be seen in the following case studies.

Part two

The projects

4

A demonstration medic farm in Libya

Introduction

The first Australian projects in the region were in Libya and they began in 1974. The overtures of the Libyan Government were welcomed in Australia because of the opportunity they provided for trade in agricultural machinery, and technical assistance was made available from within State Departments of Agriculture.

One of these projects was initiated and operated by the Jabel el Akhdar Authority in Eastern Libya and it used a demonstration farm to provide local farmers with a model to emulate. This chapter examines the way in which the demonstration farm operated and evaluates its role in persuading neighbouring farmers to adopt a medic/cereal rotation.

The Jabel el Akhdar Authority project

The Jabel el Akhdar Authority was set up as part of a policy of investment of the oil revenue that was then at its greatest. The Authority was invested with power to achieve a wholesale improvement in the facilities available to the communities within its scope. As far as farming was concerned, it was not only to improve the system being used but also to provide the resources needed by farmers to adopt new technology. The El Marj district has always been regarded as the most fertile part of Libya. The old centre of Cyranaecia provided grain for Rome and when the Italians colonised Libya it became an important part of their program of colonial settlement. Many of the farmhouses and farmbuildings remain from that period, but are unused by the new generation of Libyan farmers. By the 1970s, the fertility of the district was on a downward slope, the yield of wheat was on average little better than half a tonne/ha and the grazing, such as it was, could barely sustain one sheep/2 ha. Sheep were fed on purchased grain, straw and sparse cereal stubble and were taken to graze in the rangeland in the winter.

The concept of an average rainfall was notional only – about 300–400 mm – but the pattern of winter rainfall could vary from 150 to 570 mm and could come early or late in the season (Day, 1979, interview). Hot, dry summer winds are common, as are occasional winter frosts.

Land reform and the reorganisation of the farms

Under the land reform program carried out by the Gadaffi government, an allocation of farmland took place. Because the fertile strip of the cereal zone tails off to rangeland to the South and to the West of El Marj, the size of farms was to become more extensive as they bit into the more arid rangeland. Eventually the plan was to have private strips of wadi bed forage being used in conjunction with common rangeland, such as that envisaged for the Wadi Karoubeh sub-project. By the time the Authority took over, the cereal zone of the Jabel el Akhdar project consisted of about 2000 private family farms of 80–100 ha each from within that area of 160 000 ha of fertile land that ran roughly along a 200 km stretch of the Mediterranean coast. Each farm family in the El Marj district was provided with an 80 ha farm complete with a substantial farmhouse and outbuildings, a combine seeder, scarifier, harrows, water tank, 80 h.p tractor, a flock of 100 sheep, a perimeter cyclone fence for each farm, as well as seed supplies, a beehive, fruit trees, a small poultry flock, and the means to irrigate the fruit trees and a small vegetable garden (Anon, 1978, p. 99).

The details of the reorganisation of land ownership of the area under the Authority remain unclear. Benkhail and Bukechiem in their study of the region say that when land reform did take place the authorities took care to allocate land to those who either traditionally farmed or grazed the land or were landlords (Benkhail & Bukechiem, 1989, p. 75). We have been told by those who took over the land that before the re-allocation many of them owned varying sized portions of land and had used it to grow wheat after wheat both for private use and for commercial sale. We were told that both the landowner's sheep and those of nomadic flockowners were grazed on the stubble from these crops at the end of summer. Conversations with Libyan farmers both in the El Marj region and out on the steppe led us to believe that the landowner took his sheep to the rangeland in winter. It may well be that landowners had family (or at least tribal) links with those whose flocks came from outside to graze in summer and that family or tribal links also made possible the landowners' reciprocal use of rangeland grazing at other periods. There seems to have been little resentment among the original landowners at the re-allocation of the land into regimented and equal lots. This may be because the new owners were for the most part the old owners, and because the housing and equipment that went with the new allocation was lavish and valuable.

The farms were not given to each family – the new owners were debited with the capital cost involved, but repayments were scheduled on a basis that was not oppressive and only one third of the capital cost had to be repaid. No farmer, however, was allowed to employ another to do his work and any farmer 'renting' or 'leasing' his farm to another was swiftly called to account. He either resumed the work himself or the farm was re-allocated to another more willing worker (Personal communication from Bashir Joudeh, noted in 1979; and Anon, 1978). This primary policy may still exist, but some changes have occurred due to a concern expressed at the People's Committee on Agriculture Conference in 1982 that the farmers in the Jabel el Akhdar were developing into successful capitalists and also to monetary and fiscal policies imposed by the Gadaffi government as the country's oil wealth declined and tight controls were placed over the disbursement of cash by farmers and others (Chatterton & Chatterton, 1982).

The link with Australia

The concepts underlying this project reflected a remarkable and innovative approach to agricultural development. Bashir Joudeh who, originally as Chairman of the Authority and later as Libyan Minister of Agriculture, designed the project, negotiated the contracts, and put it into operation, was a graduate of the faculty of Agricultural Science at the University of Western Australia. He had gone there to study dryland farming in the early 1970s and had made it his business to go out onto farms in the countryside and talk to farmers. What he saw and heard convinced him that the medic/cereal rotation being operated by Australian dryland farmers would fit conditions which Libyan farmers faced (Personal communication).

The project plan

The initial project consisted of four parts to be undertaken simultaneously:

- demonstration and research,
- extension and on-farm training,

- rangeland development and community participation,
- provision of appropriate machinery and equipment.

(a) Demonstration and research

Bashir Joudeh wanted to set up a working farm to demonstrate the medic/cereal rotation to his farmers. He wanted farmers to see a demonstration of shallow cultivation using scarifiers, precision seeding using a combine seeder, regenerating medic pasture, sheep relying for their nourishment on medic and the cereal stubble from crops benefiting from the nitrogen made available from the pasture and fertilised with superphosphate. He set aside about 1000 ha of the project area for this farm of which a portion was to be used for trials or other experimental work that might be required to solve a particular problem such as a pest or disease, or difficulties in finding compatible medic cultivars.

(b) Extension and on-farm training

Having established a farm on which to demonstrate the 'ideal' he wanted to ensure that the local farmers on their own farms would be taught how to emulate it.

He did not attempt to rely on specially trained Libyan technicians to extend the knowledge demonstrated on the farm to his farmers, but instead decided to bring in expert medic farmers who had extensive experience of operating and managing the system on their own farms, employing them to transfer that knowledge to his farmers. The contract with Seedco (South Australian Seedgrowers' Cooperative) for the medic seed which was imported from Australia included the provision of groups of expert medic farmers to do this.

(c) The rangeland and community involvement

In addition to the relatively fertile farmland centred around El Marj, Joudeh was responsible for a huge area of rangeland covering the Benghazi Plain and the regions south of the Jabel el Akhdar. His ideas for improving that also proved to be radical. Instead of going down the conventional path of fencing off the range, excluding stock and allowing natural revegetation to take place while planting saltbush (*atriplex* spp.) and other fodder shrubs, he decided to sow medic directly onto the rangeland and to see if he could rapidly develop a rangeland pasture. He was confident that the medic would thrive in arid rangeland conditions (particularly when broadcast with superphosphate fertiliser) and asked the Seedco farmers to carry out a sowing program on the Benghazi Plain.

He set up a small pilot project at Wadi Karoubeh and asked the Seedco farmers to sow medic (dressed with superphosphate) in the wadi bed to see if the natural seasonal inundations would provide sufficient water to enable it to grow abundantly.

At the same time he understood the need for rangeland communities to develop management regimes for the improved pasture and the local livestock owners were closely involved from the start.

Saltbush shrubs did, however, creep into the project and nursery-raised plants were planted and watered by hand on a part of the vast limestone plateaux on either side of the wadi bed.

(d) The provision of appropriate machinery and equipment

In order to facilitate the adoption of a medic/cereal rotation, each farmer was equipped with a scarifier, harrows, a combine seeder, and a tractor so that he could begin shallow cultivation and precision seeding and fertiliser placement as soon as he was shown how to operate the implements.

The rationale behind the project

From his own experience and observation in Australia, Joudeh was convinced that the climate and conditions of Libya were so similar to those of southern Australia that it was worth the gamble to go in and sow the medic and build a farming system on the results (Personal communication). He was not foolhardy. He set up the demonstration farm not only as a living illustration of the way in which a cereal/livestock enterprise could be run using medic pastures as a base, but by including technical staff in its team, he was ensuring that if research was needed in the initial stage it would be available. In effect, what he was doing was setting up a duplicate of the research farms like the South Australian Roseworthy College and Minippa Research Farms.

There did not exist in Libya at the time a well established and proficient farm extension service, and Joudeh's radical method of providing farm advice avoided having to find funds to pump into an institutionalised service with its administrative hierarchies, career paths, educational facilities and training programs.

By providing the appropriate machinery for shallow cultivation and precision seeding and by employing experienced medic farmers to show

Libyan farmers how to set up and use the equipment and how to maintain it in good condition, he showed that he understood that farmers could not successfully change their cultivation practises if they continued to use inappropriate equipment.

The rangeland aspect of the project was truly pioneering. In South Australia medic pastures were being used effectively on the far West Coast and the Murray Mallee where average rainfall is 250 mm. Joudeh wanted to extend the medic pastures on to the Benghazi Plain, which has an average rainfall in the vicinity of 150 mm and he wanted to do it by untested techniques such as broadcasting seed and fertiliser from planes. He also planned to sow large areas on the ground using the wide-scale scarifiers and combine seeders manufactured in Australia for farmers who needed to sow up to 2000 ha as quickly as possible. These machines were used to rapidly prepare a shallow seed bed on the uncleared plain, sow the medic seed and place the fertiliser all in one operation. He employed a young Libyan agronomist to evolve with the Wadi Karoubah community a means of developing an acceptable management regime that would utilise the medic pasture and rangeland vegetation. No work had been done in Southern Australia on rangeland as arid as this, but Joudeh was not prepared to wait until research had prepared the path for him.

He may have considered that the cost of his program would be no more, and probably less, than a long term painstaking research program to test and evaluate suitable medic cultivars and to develop specific techniques for ground preparation and sowing. His approach was very like that of the South Australian farmers who evolved the medic/cereal rotation.

Acquiring the equipment and the expertise

Normally agricultural projects in developing countries tend to depend on the guidance and supervision of international agencies such as FAO for their choice of project personnel and equipment, but in this case Bashir Joudeh himself approached the experts he wished to employ.

He went first to a machinery firm in Australia because he was impressed with the efficiency and robust nature of the Australian scarifier and the combine seeder and its ability to carry out the seed bed preparation, seed and fertiliser placement and covering operations in one passage over the ground. He realised that it was not enough simply to buy components, but that skilled personnel were necessary as well to show farmers how to use and maintain implements and machines. So he procured not only the appropriate machinery for a medic/cereal system but skilled mechanics as well to set up the machines and train Libyan counterparts in their maintenance and use.

He imported the lightweight cyclone fences used on farms in South Australia to keep sheep in paddocks and brought in a fencing expert to help erect them on Libyan farms.

In the same way he went to the South Australian Seedgrowers' Cooperative (Seedco) not only for the medic seed but also for expert medic farmers. Finally he went to the South Australian Department of Agriculture for technical expertise.

He received immediate responses from the machinery manufacturers, the fencing firms and Seedco. These commercial interests were enthusiastic about government policy to encourage trade and were irritated when the Department of Agriculture simply put the Libyan proposal in a 'pending' basket. The government had offered to act as an umbrella under which commercial interests could enter these 'difficult' countries, and the procrastination of the Department of Agriculture was hardly playing the game. A protest was made and a direction given that an appropriate official be sent to Libya to undertake the negotiations that resulted in a technical team being sent to El Marj to set up the demonstration farm.

The contract and negotiations for the demonstration farm

The purpose of the farm was to show an 'ideal' and to provide an isolated site where problems could be staged and solved. But this demonstration was not to pre-date the transfer of the same knowledge to private farms. Both were planned to occur simultaneously.

The contract for the demonstration farm (signed by the South Australian Premier on 10 June 1974) made it clear that the farm was to be 'developed as a cereal livestock mixed farm to be managed and operated in accordance with the practices which have been shown to be effective on similar areas in South Australia' (Agreement, 1974, p. 1).

The Appendix attached, which contained details of staff required, provided leeway for the South Australians to send an expert farmer as Officer-In-Charge to be responsible for 'overall management of the Demonstration Farm, including financial budgeting, farm development, crop and livestock programmes, staff supervision, and the development of experimental programmes' (Agreement, 1974, Appendix (1), p. 2).

The majority of the skills required were of a kind possessed by most experienced South Australian medic/cereal farmers. The South Australian Department of Agriculture insisted that the team leader and the majority of the staff be recruited from among technically trained officers of the Department.

Two farm assistants ('preferably experienced young farmers from the South Australian Wheat–Sheep zone') were required to 'conduct the field operations associated with the crop and livestock programmes and with the experimental programmes' (Agreement, 1974, Appendix (1), p. 3) and the Department's officials agreed that farmers be recruited for these positions, which they perceived as being the equivalent of the farm managers or foremen employed on their own research farms in South Australia. These two 'specialists' were to be paid half the salary of the technical specialists who would conduct the 'experiments' (Agreement, 1974, Appendix (2), p. 1).

Later attempts to improve the status of the experienced farmers by classifying them within a category of the South Australian Public Service in a way that would entitle them to salaries and allowances more suited to their skills and experience, were strongly resisted by the technically qualified Department of Agriculture officials who were asked to implement the proposals. They were outraged that farming should be considered of an equivalent value (in monetary and allowance terms) to their diplomas and bureaucratic skills (Chatterton, B.A., Personal Diary, 1979).

The demonstration farm in operation

Before the technical staff arrived, the Seedco farmers (under the terms of their separate contract) sowed the demonstration farm with cereal crops and medic pastures using shallow cultivation and a combine seeder. This was the first gamble and it paid off handsomely. The season was a reasonable one and the records show that oat/medic hay was cut from 70 ha and averaged 1.6 tonnes/ha, and that a flock of 1055 Barbary ewes was grazed on the cereal stubble and the medic pastures. Wheat yielded (depending on variety) up to 2.3 tonnes/ha. The average was 1.4 tonnes/ha over the 500 ha sown.

As far as the Australian farmers could judge the neighbouring farms reaped the normal Libyan average of about 0.5 tonne/ha.

All cereals on the demonstration farm were sown with 180 kg/ha of superphosphate, which was quite heavy, but because the soil appeared to be deficient in phosphate a larger quantity than normal was used.

The official records begun by the technical staff after they arrived only mention superphosphate being used, but the Team Leader later recalled that in fact some nitrogen was used because as no medic had been previously established it was thought that some nitrogen would be necessary to ensure a good yield of cereals (Day, 1979).

The Seedco farmers planted half of the farm to medic using seven cultivars of commercially available medic and two of sub-clover and although they experienced some damage from grubs, sitona weevil and lucerne flea, the production and seed set was good for four of the medic cultivars – Jemalong (*M. truncatula*), Snail (*M. scutellata*), Borung (*M. truncatula*), and Paragosa (*M. rugosa*).

No bare fallow was used before the cereal crop. Once the initial medic pastures were established and were regenerating, cereals were sown in direct rotation with the pasture and no nitrogen fertiliser was applied. An area of continuous wheat was kept as a demonstration.

Neighbouring farmers cultivated the land with a three furrow disc plough after the opening rain, sowed by hand, and then turned the seed in with the same plough. All land was sown, there was no fallow. Wheat followed wheat unless there was a severe drought.

Bashir Joudeh recalled in 1979 the cynicism with which the local farmers watched the shallow cultivation being carried out by the Australian team. The 'scratch' depth of the scarifier and the dividing of the land into two – one of which was planted to pasture – caused the farmers to tell Joudeh that he had made a mistake in calling in these foolish foreign experts. The results of the first year provided a vindication for Joudeh (Personal communication, 1979).

The technical team and the two farm managers from South Australia operated the farm in the second season (August 1975/July 1976), which was another good one in terms of total rainfall, although patchily distributed.

Good cereal yields were again recorded and the medic pastures and cereal stubble fed a newly imported flock of 660 Corriedale and Suffolk sheep as well as 32 Shorthorn cattle from Australia. In addition the farm had to take in and feed an extra 3300 local sheep in summer and, later, an unexpected influx of 16 000 sheep which they were asked to hold on the farm until they could be distributed to local farmers.

In the third year there was a drought. Only 58% of the average rainfall fell and nearly one third of this fell in April, long after it could be of use to the medic and cereals. Because of the drought, none of the cereals planted were reaped for grain. The medic was, however, able to produce 'appreciable quantities of seed' (McPhee, 1980, p. 9). The sheep appeared to survive the drought well – the Barbary sheep producing 83% lambing and a nine-month wool clip of 2.8 kg/hd. The team had imported a small flock of imported Corriedale/Suffolk sheep but they did not lamb due to the fact that they could not join with the local sheep and their health was suffering as they were not particularly suitable for hot dry conditions, and were vulnerable to local diseases. In addition to normal farm work, cattle yards were built, contour banks for soil conservation were constructed both on the demonstration farm and on private farms, medic was planted on some private farms and several field days were held.

The fourth season was excellent with 157% of average rainfall which fell at just the right times. To the great relief of the demonstration farm team the medic pastures regenerated abundantly and there was no need to re-sow. The Corriedale/Suffolk sheep were discarded 'as they were not adapting to local conditions' (many had, in fact, died) but the medic not only supplied grazing for 1800 sheep and the cattle but also produced 10 325 bales of hay (some of this hay was medic/oats) – with an average yield of 2.4 tonnes/ha. The discarded Corriedale flock were replaced with a Barbary flock but these were also 'discarded' in 1979 'because of management problems'.

Episodes of fence cutting and pasture poaching in the early period of the farm disturbed the team and they decided not to persist in the attempt to establish a farm flock to demonstrate the normal livestock phase of the medic/cereal system. The medic pastures were therefore used to provide large quantities of hay, agistment (an arrangement whereby pasture is rented to livestock owners) for large flocks of sheep which were brought in *ad hoc*, as well as feed for a herd of imported cattle established briefly by the technical team.

In the final year of the project (1980) an excellent season again ensured that the 900 ha produced very good cereal yields and sufficient medic pasture to provide agistment for 8000 local sheep. The pastures also supported a new herd of cattle (a mixture of Jersey, local, and Shorthorn crosses) numbering 145. As well as providing this amount of grazing, 10 720 bales of hay were made (McPhee, 1980).

Research and extension

The technical team tended to concentrate on recording trial results from sowings of varieties of wheat and barley and oats, different applications of herbicide for weed control, experiments to ascertain optimum sowing time for cereals, the effects of different fertiliser and seeding rates and so on, and the collection of statistics for future farm planning programs. They also spent time surveying fields in order to make contour banks to help resist erosion.

Assistance was given to an FAO funded medic seed collection project undertaken by Dr G. Gintzburger and subsequent sorting revealed over 1000 varieties of medic indigenous to that section of Libya surveyed. From this collection, 560 lines were selected and an evaluation trial was conducted at Benghazi plains from which eighty of the most promising lines were selected for further testing.

Field days were held on the farm for local farmers, and the team prepared a Farmer's Work Diary and calendar (McPhee, 1980, p. 13). Altogether, 431 farmers attended the field days and Beida University students made a visit. A video film was made of the demonstration farm during the cereal harvest and this was later shown with great effect to important people and visitors. The technical staff did not believe it was their responsibility to carry out face-to-face extension work and the Seedco farmers carried out the bulk of the farm advisory work (McPhee, 1980, *op.cit.*).

An economic comparison between the medic and the traditional system

A useful piece of work undertaken was an economic comparison of the medic *vis-à-vis* the traditional system put together in 1979 by the agricultural economist employed in the farm team. A comparison was made of two 80 ha farms, both of which had 70 ha of arable land.

- The tillage system used for the traditional system consisted of deep ploughing and no fertiliser or herbicides. The cost to the farmer of this tillage program was LD 36/ha.
- The cost to the farmer growing his cereal after a medic pasture was LD 58/ha if he used both superphosphate and herbicide.
- Using the traditional system cereal yields were 0.8 tonnes/ha while on the farm using a medic/cereal rotation the yield was 1.8 tonnes/ha. The difference in yield more than compensated for the increased cost incurred.
- Using the traditional system (feeding sheep in the steppe in winter and sharing cereal stubbles and roadside volunteer pasture with nomadic flocks in summer) the farmer could maintain only 50 ewes. This flock would produce 30 lambs and 100 kg of wool.
- By contrast, the farmer who had 35 ha of his arable area in medic pasture could feed 105 ewes all year round using the cereal stubble as late summer grazing when it became available on the other half of the arable area. In addition to the grazing he could use about 10 ha for medic hay which he could sell for cash, and he could expect to get 74 lambs and 210 kg wool from his 105 ewes.

- Costs associated with the medic pasture were calculated at LD 5.5/ha, including the application of superphosphate to encourage the regeneration of the pasture each year and allowing for about 20% of pasture to fail to establish and renovation of the remainder every twelve years.

This information is presented in Table 4.1 (sourced from Prance in Chatterton & Chatterton, 1989, pp. 72–3).

Although this comparison provides a simple illustration of the benefits of medic, it was not used as the basis of any farm extension program. Our decision to publish it in our own report of the project in 1979 was deplored by officials in Australia who did not consider it sufficiently 'scientific'.

The final report of the technical team

In 1980 the technical staff and the farm managers left the farm and the contract came to an end. A change of government in South Australia led to a cooling of the relationship between the two countries and Joudeh made the best of it by asking the Seedco farmers to take on what was left of the farm after cutting it into portions so that several more Libyan farmers could be allocated land.

The departing technical staff prepared a list of 'Recommendations for the Future' full of ideas and suggestions. They were proud of the buildings and amenities that they had erected, and of the results they had achieved. It had shown Libyan farmers a formidable production unit based on medic pastures.

As the farm was to be cut up and distributed among local farmers, the team suggested that about 70 ha should be kept for a demonstration area and that a portion be used as a 'control' (that is, continuous wheat on wheat) to remind farmers of the difference between the traditional and the new systems, and that the rest of the farm, until it was distributed, should be run as a wheat/medic farm.

The team were also keen on the idea of allocating a portion to the production of seed wheat, but were still a little puzzled as to what to do with the prolific medic pasture. Perhaps it could be cut for hay and used in a local dairy? Another suggestion was that medic hay could be produced and used as supplementary feed through the summer for a livestock production unit.

Many of the team were keen to see the farm remain a centre for research for example, to continue the program of establishing guidelines for 'time of cereal seeding'. They warned that '*immediate*' research should be undertaken into dealing with two insects noticed on the demonstration farm and

Table 4.1. *An economic comparison between existing and medic systems in El Marj, Libya*

	Gross Income	
	Medic/Cereal (50% cereal/50% medic)	Conventional (100% cereal)
Wheat	6300 LD	5600 LD
Sheep	3087 LD	1295 LD
Hay (sold)	738 LD	Nil
Total	**10 125 LD**	**6895 LD**
When costs of the respective systems were taken into account the result was as follows:		
	Medic/Cereal	Conventional
profit per farm	**7372 LD**	**4050 LD**

wanted '*priorities*' to be given to agronomic research into time of seeding and more on rates of seeding and fertiliser for cereals; a token 'selection of more suitable medic varieties . . . particularly for the lower rainfall areas' should be undertaken; and, it would be 'beneficial' to have a 'comprehensive sheep production research programme investigating fertility, lambing times, growth rates and rations . . .'. A more useful suggestion was that 'As the successful ley farming techniques have now been adequately demonstrated at the Demonstration Farm the next phase is in extension', the farm could be used to train Libyan technicians as area managers.

The team, who had never felt themselves officially responsible for face-to-face farmer extension advice but felt that it was an important next stage, noted that 'the beginnings of this (extension) programme have been successful with the annual visit of farmers supplied by Seedgrowers Coop who have assisted with seeding each year'. No mention was made of the economic comparison carried out in 1979 and its potential value as part of an information program designed to persuade farmers to adopt a medic/cereal rotation (McPhee, 1980, pp. 19–24).

Personal points of view

Two of the personnel employed on the farm (one a technical expert and the other an expert farmer) later provided personal accounts of their experience. These accounts fill out the project reports and provide a window into the often unreported factors that affect projects and determine their outcome.

(a) The first team leader

The original team leader (agronomist Henry Day, who had been the first Department of Agriculture officer to adopt the farmers' system of regenerating medic at Minippa Research Station in 1953) sketched in an interview his impressions of the demonstration farm and its performance.

He had continued to visit the farm until the conclusion of the contract and overall he was proud of the result. The medics had done well and had proved their adaptability to the dryland farms of the project zone.

His first impression of El Marj had been that it was 'overgrazed and overcropped'. There were no 'farms with fences as we understand them' – just 'local people growing cereals' and 'a movement of nomadic people moving up from the desert after the harvest' to graze the stubble.

The splendid rhetoric of the day required that the team 'establish the climatic factors and the social implications that may affect the system of farming to be adopted' and then develop 'a system of farming . . . based on the South Australian system and one which we thought would suit the environment and the social situation'.

The team responded to the first requirement by obtaining information from 'local records, and FAO and the World Bank and from other organisations that had been in the area and were able to come up with rainfall patterns and intensities'.

The social context was largely ignored. No strenuous efforts were made to find out what it had been and there seemed no reason to do so. The team took for granted that the 'social guidelines . . . came out of the change in social structure that the Libyan authorities were aiming to make in the area'.

Henry Day doubted that

> they [the surrounding farmers and those whose grazing rights had been disrupted by the establishment of the demonstration farm] really resented our coming, but we did experience some problems with lifting and cutting of fences but this was, we gathered, principally because the nomadic people moving through could see that over the fence there was a lot more feed than there was outside the fence and considered that they should have a part of that too

With regard to training of farmers he said

> I doubt whether [direct training in farm operations on the demonstration farm] is feasible. I think probably our aim should be, and it will in fact be, to try to get Libyan farmers to come onto the Australian farm in small groups to look at one specific point at one time. It may be the seeding operation, or the cultivation operation. But just one point at a time to look at it in detail.

He regretted that when the technical team left Libya they left behind no extension agents trained to advise farmers using medic, but there had been no Libyan counterparts available to train. Sudanese and Egyptians did the farm work and animal handling on the demonstration farm under the direction of the two Australian farmers, and the Libyan extension agency for the whole region had consisted of two persons who were in fact 'area managers' each responsible for 100 farms and with no specific brief to extend the medic message – and no experience of it themselves even if they had been given such a responsibility (Day, 1979).

(b) An expert farmer

The two farm managers (both expert medic farmers) threw a different light on the conditions under which they all worked. They described their position in similar terms – that of 'underpaid foreman' and they resented being treated as inferiors by young men with very little if any farming experience and only a diploma and possibly some experience in a district office to buttress their claims to superiority. However, their resentment did not affect their enjoyment of their task nor their keen eye for the bizarre.

One of the farm managers employed on the project was Bill Kelly, a not very young, but a very experienced, farmer from Kangaroo Island in South Australia who had spent time in the Benghazi–Tripoli region during the Second World War. During the fighting he was captured and spent time devising several hair-raising escape plans – the last of which was successful.

He wrote regularly to his family from January 1975 just after he arrived at the El Marj farm.

On [the farm] we [that is, the project team] have virtual control and are supposed to run it as we would an Australian farm in the most efficient manner available to us. By next year we should have a complete set of tools, implements and machinery of our own, ordered to our own specifications. These we have already placed on order and I would say they would be fairly typical of what you would use on a mid-north farm. [This referred to a typical dryland South Australian farm in the 300–400 mm rainfall zone producing both sheep and cereals.] We haven't overdone it and have gone for a Shearer PTO header instead of an Auto one as we think that is sensible (Kelly, 1975–7).

He describes the growth of the medic pastures that had been planted in the previous autumn on the demonstration farm by Seedco farmers and expects that

with a bit of warm weather they should be a picture . . . We are [he went on] due to order some Australian sheep any time now. They [that is, the technical team] were inclined to go for Corriedales but I am pushing for comeback-ewes mated to

Merinos. It is real Merino country, no doubt about it and the nearer they get to
Merinos the better. They are talking of flying a plane load out.

He had already inspected the medic program being carried out on
rangeland on the Benghazi plain which stretches

from the sea to the escarpment about 20 miles inland and is some 200 miles long.
About 100 years ago it was all scrub with a certain amount of pasture . . . now it is
mostly short limestone due to overstocking and wind erosion . . . We had a day with
the boss (Libyan) of it last Wednesday and he knew every grass and shrub on it. On
the bits with soil they sow medics and grasses and trees, mostly wattles . . . there are
some very interesting medics on that plain

he added reflectively.

On about February 9 the team had their first experience with local sheep.
The medics he began 'have really shot ahead . . . the best are Paragosa and
Jemalong . . . the Snail looks spectacular too . . . almost 1000 acres (400 ha.)
. . . It makes everything else around this area look very second class, and
people are coming to look at it from all over'.

But the sheep . . .

There are 400 of them and the roughest lot of mongrels you would ever see. All fat
tailed barbarys and all colours of the rainbow. Not much wool on them but what
there is is hairy and matted. They were mated ewes without their 2 teeth and cost
(Aus)$100 each. Great panic when they arrived. We had a small paddock ready for
them but we had to shed them . . . on account of roaming dogs . . . we were advised
not to let them out for a feed on our pasture or they would die of bloat . . . A lot of
Cock of course and we let them out with no ill effect and they look a whole lot better
for it. We will be inoculating for enterotoxemia tomorrow and that should see them
right. But oh curses having to shed them each night. How I long for a good
Gallagher electric fence.

He paid the sheep a tribute.

One thing about these sheep is that they haven't bred all the brains out of them like
we have ours and they have been wandering around with shepherds for generations,
so when we shed them one of us goes inside and makes a noise with his tongue like a
kid imitating a motor car and lo and behold they all start bleating and walking after
you.

However, the arrival of the sheep is 'making a muck of a lot of our plans'
. . . 'We had three paddocks ready for them and can't put them out there as
it is too far for them to come back to be shedded'.

The directness of the Australian farmer was to the fore in finding the
solution. 'We have ordered a big roll of that mesh . . . and may have to put
them up in the paddock to hold them at night. We have also ordered a lot of
strychnine. I wish I could get hold of a decent .303 [rifle] and a few rounds.

Those dogs really need thinning out'. He did eventually get his way with the dogs, but that is another story.

In mid-April

We are having a devil of a time with the sheep ... So far the dogs have only got a couple what with yarding them up and the Sudanese watching them at night. But now they are really loading us up. We have 4,500 on the place and more to arrive. These are allegedly for farmers who are to be allocated their blocks ... the main problem is that we have no water at all on our 1000 acres.

Once again the Australian farmer's innovative genius came to the rescue.

'We dug up some old Italian underground tanks with cement troughs attached and we rigged a centrifugal pump on the tank and pump the water out when we water them. The tank holds about 40,000 gallons and we have to keep filling that with a big Fiat water tanker'.

He notes that

My respect for these sheep grows daily. If you put one of our mobs [Australian sheep] when thirsty on a set up like that they would muck it and only half of them would get a drink. But the Sudanese bring them to within 100 yards and let them in in batches of about 70. The rest stay put until these have drunk and go out the other side.

At the end of summer 1975 (August 3) he writes that there has been no rain since April and that the local sheep flocks are all hungry yet

on our farm our sheep are mud fat and we have feed to burn. In fact we have about another 3000 sheep in on agistment (no pay) just to get the stubbles down enough to get an implement over them. The medics have thrown a power of feed. It really is ridiculous to see the difference between our stubbles and the local ones without medics. If only we can get the message through ... we really will have done something. That is really the whole object of the exercise.

Of course, the lush feed in the demonstration farm's 1000 acres led to trouble. 'The locals take a dim view of the fences and cut them at night and put the sheep in at night which is most annoying but understandable if they have starving stock'.

Kelly yearned for the Gallagher electric fence but the vested interests of aid programs interfered. Bashir Joudeh (who was the Development Authority Chairman at that period) told him 'that someone in England was sending one to him by air as a present'. When it came it was a 'toy' and Kelly vowed not to have anything to do with it. A colleague was 'playing around with it, but' (wrote Kelly sourly) 'all it will do will be to give electric fencing a bad name'. Kelly, early in the piece, saw the aid industry in action.

He records that Joudeh (who was prepared to accept a piece of equipment as a present even if it was not the best piece of equipment for the job) was not so naive when it came to the big projects. He diverted the Australian soils expert on the farm to examine two proposals in detail and provide him with a report on their suitability to the Jabel el Akhdar. They were only two of many feasibility studies put up constantly by the Ministry of Agriculture on behalf of, according to Kelly, 'a world racket of people of little knowledge who rush about the world knocking up expensive feasibility projects for developing nations, few of which could even work'. Both proposals, costing well over US$50 million each, were rejected by Joudeh.

Sheep galore In January 1976 the problem of sheep on the demonstration farm turned into something of a nightmare. An irrigation scheme down in the desert to produce lucerne for large flocks of sheep went wrong. The sheep became bloated on the lucerne (alfalfa) and had to be evacuated out. Of course, there was no feed anywhere in Libya except on the 'Australian farm', so the sheep were trucked, dying and hungry to El Marj. Each truck took four days to travel the distance and each truck carried 65 sheep. The Australians were horrified. They were told to expect 10 000 sheep and prepared six paddocks for them. But when the 10 000 had arrived in a bizarre convoy they were told that was only half the flock and a further 10 000 was still to arrive. The Australians buried the dead with bulldozers and then decided to shift about two thirds of the remainder down to a depot five miles away where they could be fed with straw and hay.

Kelly recorded that 'We expected a lot of trouble [in walking them down] but the leaders just kept following a landrover with hay on it and they were strung out for miles'. That left the farm with 3000 sheep and the team felt this would be enough to continue with (they also had on hand their own flock of 1000 ewes that had lambed, producing about 800 lambs that also required feeding and care). However, at the depot things went from bad to worse – they hay provided was in reality poor quality straw and when that ran out the sheep were fed barley out of bags direct onto the ground. When rain came the result was catastrophic. Those sheep that survived had to remain until March as there was nowhere else for them to go.

Kelly calculated that if one gave a notional value of $100 to each sheep lost in the exercise it had been a costly experience for the Libyan Government. The money spent on the misguided irrigated lucerne patch deep in the desert could have paid for many hectares of good medic pastures to be established on farms in the cereal zone.

Lack of contact with local farmers The good news for the year was that the lambs and ewes on the farm were no longer being attacked by dogs. The Gallagher electric fence had arrived with Christmas. Nonetheless, Kelly felt dissatisfied with the general result of the farm – not the productivity, but 'I feel the disappointing thing has been the lack of cooperation with the ordinary Libyan farmer'.

He drew the contrast with the demonstration farm behind its fences and the Seedco farmers sharing bed and board and daily farm work with the local farmers – 'I feel' he wrote 'that the seedgrowers did a better job in this regard than we have so far'. He had no doubt that the local farmers would see the sense in medic. 'I have never yet seen a farmer in any country who couldn't realise that if there was more money to be made by adopting new methods he shouldn't quickly follow suit'. Yet, given the reluctance of the project's technical team to get too closely involved with individual farmers, he had to pin his hopes of better adoption of the system on plans 'to hold demonstration days ... that should create a lot of interest'.

Problems with ill-chosen imported sheep The dilemma over the attempts to establish a permanent flock of sheep on the medic on the farm continued and in September 1976 Kelly wrote that the harvest was over but that the 'sheep have been a real problem. The Corriedales [imported by plane from Australia against his advice] appear to get all the diseases about the place, from worms to pleuro pneumonia'. He believed that the nightshedding was a major cause of the problem 'If we could leave them out all night there would not be such a disease problem. They are also susceptible to cattle tick in spite of frequent dipping. The Suffolk rams [the other part of the misguided program] get worse tick ... because of them having more bare areas and they really look pretty awful'. To add to these problems the imported sheep 'don't appear to mate with the locals' and in spite of many strategies to bring this about, Kelly recorded in December that no Corriedale ewes produced lambs.

A clash of philosophies It was the acceptance by the technical staff of the haphazard manner in which the medic pastures and cereal stubble were exploited on the demonstration farm that revealed the huge gap between the philosophy that motivated the technical team who were satisfied with the demonstration of productivity in itself and that of the farmers who required that the productivity be translated into a management regime suitable to the local farms.

Kelly understood this and provided an illustration of the contrast when he wrote in September after the summer harvest of cereals

We need a lot of stock on the place at this time of the year to clean up the stubbles enough to get an implement through them. We don't like to burn the stubbles (as they would do in Australia in abundant years) partly because of it being bad public relations when there are thousands of starving stock in the countryside.

Joudeh had promised to truck in another 10 000 sheep from the desert to utilise the abundant pastures, but they had not arrived so Kelly and his farmer colleague persuaded him to allow some local sheep in. 'You must understand the local tradition with stock movement' he reminded his correspondents. 'When the rains cut out in May thousands of local sheep converge on this area from south of the desert where they spend the winter. They are given access to the stubbles (on all the local farms) during the dry period and clean them up like a fallow ground.'

Of course, on the demonstration farm it was different.

We alone have them fenced out as we try and control a couple of paddocks for the break in the season when there is virtually nothing left outside . . . a few people lift the fences or cut them but this is a minor problem . . . But when we decided to let some in things took a different turn. Farmers turned up with an authorisation to say that they could have, say, 500 sheep in. Then the word got round and it was open slather. Instead of bringing in 500 it would be 700 plus 100 goats and a few cattle. Others came in with no authorisation at all.

Kelly was remarkably calm about this. 'It is all OK as long as they stick to their allotted paddocks, but when they got onto our reserve feed we got a bit niggly'.

The other Australian farmer (Sam Pfeiffer) employed on the same terms as Bill Kelly decided to go out at night and see if he could stop the poaching. After a great amount of excitement and effort he and his team of Sudanese impounded about 800 sheep and two flocks of goats. Joudeh's instruction was to charge $3 per head for their release as a lesson to the landowners, but the team, feeling the burden of being foreigners, settled for 75 cents per head and even turned a blind eye when the culpable farmer claimed that the sheep flock only numbered 500. They did not respond to his claim for compensation for a goat that died in the yard – simply 'let him off paying the 75 cents due on that one'.

At the height of the exercise of allowing local sheep in, the farm carried 10 000 sheep belonging to local farmers plus 2600 sheep 'belonging to the farm'. While Sam Pfeiffer continued to be irritated by the tendency of the local farmers to bend the rules, Bill Kelly philosophically reminded him that 'it is educating the locals on just how much feed there is in the medic stubbles. And sure enough [he writes] they don't take long to learn that lesson'.

A summing up He left the project in January 1977 and his final letter home written on New Year's Eve summed up his experience. In spite of a difficult season the medic was continuing to regenerate and perform well. The Australian Corriedales and Suffolks were dying off and in spite of a lot of attention, 'they just appear to be susceptible to all the diseases about the place'. The grain crops were speedily planted – 'about 1,000 acres (400 ha.) in just over a week, which startled the natives somewhat'.

While many of the locals continued to 'persist in ripping the ground up over a foot deep' ... it was ... 'pleasing to note that this year a lot more followed our example and used scarifiers ... partly due to the Shearer scarifier that has been issued to them ...'.

Bill, his fellow farmer Sam and one of the technical experts, Peter Marrett (who was a 'natural' extension agent) spent a lot of time helping local farmers to set their combine seeders and sow their crops. The local farmers' difficulties were compounded because not only had they to learn to use new machines but the instructions provided by the manufacturers were only in English. The translation and all further communication took place in 'broken Arabic, broken Italian, vivid Australian and much gesticulation'.

So popular was this assistance that farmers were dragging their combines from up to ten miles away to get the help necessary to use them properly. Many farmers still at this time preferred to stay with their annual wheat crop, but a few 'have caught on and are quite keen'.

Bill and his colleagues would always carry a bag of medic seed with them and try to persuade each farmer to try a little. There was difficulty in persuading farmers to use phosphate fertiliser. 'They come up with all the old yarns such as "If we get a dry season it will burn the crops off"'.

Joudeh, impatient with the farmers' doubts and fears, sent a pilot up in a small plane with a mixture of nitrogen and phosphate (18-46) to spread it over the whole district – but he was not popular with the project team as, not only did they not want nitrogen on the medic, but it ruined the fertiliser trials they had carefully set out for demonstration purposes.

Kelly felt that it was indisputable given the results on the farm that the system was suitable for the Libyan cereal zone, but he saw the difficulties that remained and that prevented an easy adoption by local farmers.

Europeans tend to fleece them by flogging off horrible types of sophisticated machinery. They get all sorts of bad advice. The public service is most inefficient. In the project itself very few [technically trained] people know anything about farming ... one of our biggest disappointments has been a lack of contact with local advisers. We get an odd lad now and then with a reasonable education who is sent out to learn. But in a few weeks they drift off to an office job where they don't get their hands dirty. It is the same the world over. The answer [wrote Kelly] is to stick with the local cockies [a South Australian slang word for farmers]. This we are tending to do more now that our own farm is in some sort of shape . . . but we would be able to do a lot more if we either had a smaller farm of our own or more farmer staff.

In spite of all the difficulties, he believed that the El Marj project was 'one of the very few places in the whole of the Middle East where a satisfactory method of stable agriculture is being demonstrated'. Kelly, among his other observations, noted the way in which the local sheep were biddable and could be easily controlled without fences being necessary. This was a contrast to Australian sheep who, because of lives spent in fenced paddocks only visited by men and dogs when something nasty is to happen, are undomesticated and very difficult to herd and control. This should have posed a question to those technical experts who believed that a medic system would not succeed without fences but it was not asked and later projects insisted on fences being installed.

What the demonstration farm achieved

The cereal phase

The demonstration farm provided an excellent demonstration of the cereal phase of the medic system. Shallow cultivation, precision seeding, superphosphate fertiliser used for cereals following a medic pasture out-produced local crops and the improvement in soil fertility helped sustain productivity even in periods of drought. The technical team on the farm were assiduous in the operation of the cereal program and took care to establish a 'control' plot to demonstrate the inability of a regime of continuous cereals to maintain yields even when fertiliser was used and herbicides controlled weed growth. On this trial, the continuous wheat (using fertiliser and herbicides) began well by showing a 25% advantage in the first year, but this declined steadily and by the final year (1979/80) the yield was 25% below that obtained from cereals grown after medic (McPhee, 1980, p. 21). The yields after medic on neighbouring farms and those in the other districts being assisted by Seedco farmers showed an increase of almost double compared to the average yield of 0.5–0.6 tonne on farms without medic. This showed what was possible even when the conduct of the system was less than perfect.

The farm enabled the technical staff to test many basic cultivation and seeding techniques associated with cereal production, and the effect of the use of superphosphate fertiliser on crops and they carried out a program for the selection and testing of various wheat cultivars and the identification and selection of local medic varieties.

The farm demonstrated how well-designed contour banks can provide a physical means of controlling soil erosion. Later when farm planning became part of the duties of the technical staff, they enjoyed the task of preparing the models and plans.

The grazing phase

The farm failed badly when it came to demonstrating the grazing phase of the system. The establishment of the medic pastures with the cultivars commonly used in South Australia was successful and there were no problems with regeneration of the pasture after the cereal crop, so one must ask why the entire system was not demonstrated.

The answer seems to be that the team of technical experts simply were not able to come to grips with the problems of acquiring and managing a permanent sheep flock. The Australian farmers (employed as farm managers) were keen to try but the early experience of sheep being stolen and pasture being poached was traumatic and the technical staff (or the management back in Adelaide) were not prepared to agree to a permanent flock being reassembled and kept on the farm. The farm managers did not have enough authority to establish a flock without the acquiescence of the team leader.

The death of the Corriedale and Suffolk flock (due to their lack of resistance to local diseases) simply demonstrated that importing sheep breeds without regard for local conditions is a risky venture.

Beef cattle were brought on to the demonstration farm to compensate for the discarded sheep, but the final team recommended that they be 'disposed of' (as they) 'are not part of the system being used by the local farmers' (McPhee, 1980, p. 20). In retrospect, the introduction of these cattle to the farm appears to have been a foolish diversion.

Potential problems from undergrazed medic pastures on the farm were avoided by making large amounts of hay and by the providential appearance of the large number of hungry sheep that periodically were dumped on the team by the Libyan authorities to graze the pastures. The switch to conservation of fodder on the farm at the first sign of difficulty was a typical South Australian Roseworthy Agricultural College response to abundant medic pastures and access to abundant supplies of labour, but it was not a good example to put before the Libyan farmer who would have benefited far more from learning to do as his Australian counterpart did and graze his sheep throughout the year so that he did not have to make hay.

Successive teams tended to excuse the absence of grazing programs on the farm with the 'nomads cut fences' syndrome, yet the local farmers managed to come to an accord with potential marauders.

The allocation of sheep per private farm (roughly 100 sheep/80 ha) was a good balance and the extremes of over and undergrazing could easily be avoided. The farmers could have been helped to experiment with the numbers of sheep they could carry on their 25 ha of medic and 35 ha of cereal stubble each year, to develop strategies to deal with seasonal flushes and shortages, and also to see how often and how much they needed to cut a supply of hay for a drought reserve.

The agricultural economist had theorised that about 150 breeding ewes could be kept successfully for the whole year on a 100 ha farm, including some hay production from the medic to supply supplementary feed for sheep during late summer and autumn. These economic data also indicated that farmers could expect a far larger increase in profit from the livestock component of their farms than from cereals, but they could only realise this if they were shown how to utilise their medic pastures efficiently.

Theories are not enough for farmers. Their livelihood is at stake when they change their practices and, when no demonstration is available, they are unwilling to take chances. Fortunately many farmers went ahead and grazed their pastures developing quite a good system, but they would have profited greatly from skilled and continuing advice. It was only when the farm was being handed over to the Seedco farmers at the end of the project that the recommendation was made that a grazing management program should be undertaken on the farm. By then it was too late.

A seed industry

In spite of the evident need for local supplies of medic seed to underpin future expansion of pastures, nothing was done to encourage farmers to collect pods and thresh seed from their own pastures so that they could enlarge their plantings, and no commercial seed industry was envisaged – probably due to discouragement from Seedco who were keen to keep the market they had established for imported Australian seed – yet an enormous amount of time and effort was invested in identification and selection programs for new varieties. Although the local farmers experienced no problems with the medic cultivars they had planted on their farms, they did need to have available a supply of medic seed for re-planting in the future to rejuvenate the pasture, and also to extend the area of medic in accordance with future requirements. When the importation of South

Australian seed ceased it created major problems for the Libyan medic farmers (El Akhrass, Wardeh & Sbetah, 1988, pp. 93–9).

However, the operation of cereal phase had not only been demonstrated adequately but had also been adopted by many farmers. This provided a model for the rest of the region that showed how to overcome the problems that later became associated with the system in the minds of technical and scientific experts employed on later projects.

5

The grazing phase and farmer training

Introduction

The grazing phase that eluded the demonstration farm staff in El Marj was demonstrated well, however, on the Western Australian project run by the Gefara Plains Authority. This Authority purchased the equipment and machinery needed to provide the infrastructure for the development sites and the individual farms which were eventually to be allocated to farmers from Western Australia and they also employed Western Australians to establish medic/cereal rotations on the project sites. Some of the Libyans selected as prospective farmers were expected to work with the expert Australian farmers during the development phase so that they could learn tillage and other skills and acquire management experience of a medic/cereal rotation before they took up their own farms.

The Gefara Plains cereal project

Background

The project contracted to the Western Australian Overseas Project Authority (WAOPA) by the Gefara Plains Authority covered a drier zone than that of Jabel el Akhdar project. The Gefara Plain, which runs east, south and west of Tripoli, is bounded to the south and east by a major escarpment leading to a high plateau. The cereal project consisted of 50 000 ha divided into seven sites, five of which were on the plain and two on the plateau. A central workshop and administrative centre was set up at Al Aziziah, south of Tripoli. The soils are mainly yellow or red brown calcareous sands, and the winter rainfall and hot dry summers are typical of those to which the team were accustomed in Western Australia. Heat waves in summer with temperatures up to and over 50° C and accompanying hot winds are common, and rainfall in winter of about 250 mm on the sites near the coast and about 150 mm on the two sites below the scarp proved not too daunting to the Australian team.

As well as the obvious trade potential that this project represented, the technical experts saw it as an opportunity to undertake research into the performance of medic and Arab sheep within the North African context. They could see that the Libyan conditions were similar to those in Western Australia and hoped that they may be able to enlarge the scope of productivity in their own marginal and arid regions by adapting knowledge gained on the project.

The project team

The Western Australian team consisted of three technical experts from the State Department of Agriculture – one in charge and two to carry out various research – and seven to eight Western Australian farmers who came in and out each year to do the actual development work, together with seven to eight mechanics to keep the machinery in order and operate some training schemes. Two Western Australian Department of Agricultural officers made periodic visits as medic and livestock specialists respectively. In 1978 a farm planner from Western Australia with soil conservation experience was recruited to prepare plans to subdivide the sites into private farms of about 100–300 ha each, the size to depend on the rainfall. Families of the team lived in comfortable houses in Tripoli (in a compound called Suani bin Vadim with other foreign workers) and had access to American Oil Company schools and facilities.

The Western Australian farmers were employed to supervise the project sites and to work 'generally with a Libyan agricultural college graduate to act as translator' to train twenty to thirty Libyans in skills such as tractor driving, shearing and fencing. The original labour force of some 100 Egyptians were replaced as the Libyans' skills in using the machinery became adequate.

In spite of the success in training Libyans to do farm work, the team found it difficult to get enough counterparts for technical training. There was 'an extreme shortage of people with an adequate educational background for the position they fill ... [the] unwillingness of young people to accept jobs away from the city and the demands of the army for up to five years national service training ... ' (Allen, 1979).

Cereals and medic on the project

When the Western Australians began their work in 1974, J.M. Allen recorded that 'The areas allocated to the Cereal Project were virtually uninhabited although they were being heavily grazed and partly opportunity cropped in small handsown and harvested strips'. The Gefara Plains Authority was responsible for about twenty varied projects in all and had employed 'international consultants' to allocate to the project areas of the plain 'not being farmed by resident landowners' (Allen, 1979). The sites were rough and infested with unpalatable shrubs. The team made good use of their Western Australian experience with such land and decided to go straight in and clean it up and sow it to pasture.

It took four years for the team to completely clear the land of unpalatable shrubs, to level it and to plant it to cereals and medic, but at the end of the period 'the total area of 50,000 ha of the Cereal Project has been developed to a condition suitable for machine seeding and harvesting'. They planted two cereal crops to clean up the soil and then medic pastures were sown. 'The aim is to continue a one year medic one year cereal rotation, although in the lower rainfall areas (150–175 mm) this is likely to be lengthened' (Allen, 1979).

Cereal yields improved markedly following the establishment of medic pastures. Scarifiers were used for shallow cultivation and combine seeders for sowing and fertilising. No bare fallow was used.

It was noted that the success of medic cultivars Cyprus (*M. truncatula*) on loamy soils and Harbinger (*M. Littoralis*) on sandy soils was a reflection of the distribution of naturally occurring strains of the two species in Libya.

The establishment of the cereal/medic rotation was claimed to be 'the first and only field experiments carried out under dryland conditions in western Libya' (Allen, 1979, p. 8). No acknowledgement was made of the work being done by South Australians in Eastern Libya.

The grazing phase

Although the introduction of the medic was originally intended to increase cereal yields, the effect of medic pastures on the productivity of livestock had 'as much impact on the sheep industry as on cereal production'. This success of the medic was due not least to the fact that 'the medic pastures . . . are so dramatically superior to existing pastures' (Allen, 1979). The Western Australian team seized on this success to put in place the following program.

(a) The nucleus flock

The Western Australian team (unlike the South Australian team) did not pursue the course of importing Australian sheep but, on the advice of Dr John Lightfoot of Western Australia, took the local sheep and set out to improve them by using a selection and feeding program based on a nucleus flock and medic pasture. The improved weight and growth of Barbary sheep grazed on medic pastures promised a major boost to both stocking rates and productivity of the animals. Allen quoted results from the nucleus flock in 1978 when 'ram lambs off medic pastures have reached a liveweight of 50 kg at 100-day weaning or growth rates of 0.5 kg per day' (Allen, 1979, p. 9). In other words, on sites where previously two hectares of land and imported grain and poor quality hay were needed to sustain one ewe all year round (and did very little to improve its condition), medic (with only a small amount of supplementary grain) had provided enough feed for up to three ewes per hectare while bringing about a significant improvement in output per head at the same time.

Lightfoot, the livestock consultant to the project, was cautious about the future. While 'it is already apparent that the improved quantity and quality of feed on offer has, in itself, achieved a major boost to carrying capacity and sheep production ... the true potential of the system will only be realised as appropriate livestock husbandry and breeding systems are implemented'. The breeding system was already on the way with the establishment of the nucleus flock of 250 ewes (selected from 10 000) and 25 rams.

Lightfoot wanted the stocking rate trials extended even further because while 'preliminary indications of one or two sheep per hectare demonstrate a very considerable increase in production compared with traditional unimproved rangeland ... there is a need for further expansion of this work on stocking rates to establish safe upper limits to carrying capacity at each of the project sites'.

The stocking rates on the higher rainfall sites reflected those being reported from El Marj in the Jabel el Akhdar, but it was the excitement of the results obtained with sheep on medic pastures sown on rangeland with rainfall well below 200 mm that was the notable achievement of the Western Australian project (Lightfoot, 1979, pp. 2–3).

(b) Grazing medic in the rangeland

Although the cereals grown on medic and using Australian cultivation and seeding methods were a success, the medic pastures in the rangeland drew the greater attention – particularly, according to Allen, 'the stocking rate trial on medic pasture on yellow sand at a site with an average annual rainfall of less than 200 mm. The trial has ewes at one, two and three per hectare and even at the highest rate lamb production has been excellent with only limited grain feeding'. This particular trial was at Adjulyat on rangeland where the medic continued to thrive and feed large flocks despite very poor seasons for some years. Even in 1979 Allen could record that ' . . . Harbinger medic produced adequate seed to successfully regenerate . . . when the total rainfall for the season was only 125 mm'.

Allen admitted that when the original identification team had suggested the use of medics on the project, even though they had been surprised at the widespread indigenous but unused medic they observed in the country, they 'did not anticipate that the production and persistence of medics would be as successful as it has been' (Allen, 1979).

Allen pointed out the difference between the conventional costly, long term and questionably effective FAO/UNDP approach of planting saltbush shrubs (*atriplex* spp.) to restore the rangeland pasture, and that of establishing medic pastures. He asked why the latter, so cheap, yet so rapidly productive, was being ignored by those who decided on the scope and type of rangeland projects. He graphically described one such project

In contrast to the Australian system, neighboring grazing project areas are being improved by planting perennial shrubs such as *Acacia* spp, *Atriplex* spp, or spineless cactus. They are propagated in nurseries, handplanted and watered through two summers, this is an expensive method of pasture. Preliminary results suggest that the medic pastures will carry significantly more stock than the perennial based pastures (Allen, 1979, p. 8).

Reporting to the Gefara Plains Authority in 1979, Lightfoot, the livestock consultant, said

A major achievement of the Cereals Project has been the large scale establishment of improved pastures based on annual medics. The potential of this achievement reaches far beyond the confines of the various Cereals Project sites. It provides for a system of greatly expanded pasture productivity and correspondingly increased livestock production through most of the 150 mm plus coastal belt of Northern Libya (Lightfoot, 1979, p. 1).

Dr J.A. Allan, after visiting the sites, wrote in 1989 that

well established medicago rotations have been introduced [providing] a relatively robust system of range management for the extensive marginal tracts of Libya with 150 mm or less of rainfall. The use of medics especially indigenous species, to improve pastures is proving a very important strategy in the fragile environments of the region (Allan, 1989, p. 124).

(c) Grazing rights in the rangeland

The grazing phase was not without problems. The team noted that 'There was little obvious recognition given to the plight of the nomadic flocks and in subsequent years the medic pastures of the Cereal Project have been severely pressured by them' (Allen, 1979).

Problems with nomads lifting or cutting fences to give their own hungry sheep access to these pastures on this project was dealt with in a summary fashion. The Libyan staff simply had culprits put in jail 'for a couple of days' (Chatterton & Chatterton, 1987, p. 130). There was no attempt made to involve local flockowners in the growing and grazing of the pastures to ensure that a local management regime was established to carry on the system after the experts left.

Farmer training

A more vexing problem was the difficulty in leaving behind a nucleus of farmers familiar with the management of the total system. Farmer training during the time that project was in operation tended to be piecemeal. Apart from the intention to train some potential farmers during the development phase, there was no scheme to involve farmers in the management of the total system. This was expected to take place later when the farms were allocated.

(a) Haymaking, shearing and culling for colour

In the meantime there were several programs undertaken to introduce Libyan farmers to some new management ideas. There was an attempt in the cereal zone to try to introduce what seems to have been an unwanted haymaking regime on the Libyan farmers. This reflected the Western Australian farmer's habitual use of medic hay as a drought reserve and their scorn for the low quality straw used by the Arab farmer.

'One good example of the role the [Australian] farmers have played' (wrote Allen), 'is in enticing the Libyans to cut medic hay. Traditionally they used the straw after harvest to supplement stock in the summer. When confronted with the idea of cutting green pasture there was open disapproval and confrontation between the local community and the Western Australian farmer' (Allen, 1979, p. 7). This dislike of cutting the green medic was unusual among Arab farmers. In other projects it would be the technical experts who became upset because the seeding capacity of the medic plant was often critically diminished particularly in the first year by farmers eager to profit from early cut medic hay.

The capacity for the foreign expert to attempt to transfer irrelevant skills to recipient developing countries was apparent in the inclusion of shearer training programs, although wool was a negligible part of the Libyan farm enterprise.

It was even more evident when as part of the valuable nucleus flock selection scheme sheep growing coloured wool were culled even though coloured wool was acceptable to Libyans and was a desirable variant in the pattern of woven carpets and other woollen goods. In Australia any contaminant of the white fleeces of the flock are ruthlessly culled and the keeping of coloured sheep for craft work is frowned upon by the wool industry.

The provision of farm machinery and spare parts

More successful was the training program for the use and maintenance of farm machinery. The Jabel el Akhdar Authority had shown the sense in providing the appropriate equipment to carry out shallow cultivation and precision seeding during the cereal phase, and the Gefara Plains Authority also invested heavily in the same equipment. The Western Australian team improved on this initial investment by also ensuring that a catalogue of spare parts was on hand for immediate repairs and maintenance. The team could boast that in 1979 all the original machinery supplied in 1974/5 was in good working order and that, by comparison, machinery supplied without supporting mechanical services was either wrecked or beyond repair after one year in the field (Allen, 1979).

Conduct of the project

The Western Australian project was well run and well planned, the Western Australian farmers received good salaries and there was no friction or resentment between the various groups because of 'unfair' rewards. Results were written up with sound rationalisations and scientific exactitude to underpin the conclusions. The few incursions of Australian conditioning that did escape into the project – such as shearing training and culling of coloured sheep – were not terribly important in the light of the wider benefits that were demonstrated such as the suitability of medic pastures to the climatic and economic conditions of Libya, and in particular to the rangeland. After the Western Australian project ceased to operate, much valuable documented experience and data remained.

The Western Australian project provided a model for grazing management with local sheep and for rangeland development using medic pastures for future projects. It was not a theoretical model; it was a practical model that was sustained for over a decade, and it should have encouraged new directions on farms and rangeland in other countries in the region.

However, there was little, if any, intimate connection between the existing farming community, or the nomadic livestock owners that probably would have ensured the conversion of the undoubtedly formidable results on the seven project sites into a practical farming system that would survive when the team left. This can be excused by the fact that the second phase of the project (the division of sites into small private farms) did not occur in any significant way while the Australian team were in Libya, but it is undeniable that even after ten years the new system operated in isolation from the farmers around it.

The Seedco farmer training program

The Jebel el Akhdar Authority began their farming training simultaneously with the introduction of the new farming system to the project. Expert medic farmers worked directly at the farm face teaching farming skills to farmers at the same time that the new system was being demonstrated on the Authority's farm at El Marj.

Background

The employment of South Australian Seedgrowers' Cooperative (Seedco) farmers as extension agents for the Jabel el Akhdar project was Joudeh's answer to the paucity of farm extension and training services in Libya. The South Australian demonstration farm was to show the Libyan farmers the system operating in ideal conditions and to provide a base where technical experts could deal with technical problems that may arise from time to time, but the Seedco farmers were to help the Libyan farmers translate the principles of the system to the context of their own farms (Seedco, 1988, pp. 52–7).

The Australian farmers were employed to come in during the winter season each year to assist the Libyan farmers to sow their medic and cereals. In all, eightyone Australian farmers from South Australia worked in Libya for short terms over the ten year period that the project was in operation. The skills and qualifications of these experienced farmers were different to those of the technical staff on the demonstration farm. They were farmers with up to thirty years' experience in operating a medic/cereal farming system. Their contracts were independently arrived at with the Jabel el Akhdar Authority and were apparently satisfactory to all concerned. They lived with the Libyan farm families and were not separated from the local community.

A Seedco team first went into Libya in 1973 and sowed the area set aside for the demonstration farm. Reporting to Joudeh in 1974 they were full of optimism. All the plantings (cereals and medic) had yielded well, some of the Snail medic had been overgrazed on one site, and the Libyan Ministry had themselves planted a control of cereals against one of the sites planted by Seedco and were recording the results. A Libyan agronomist had been appointed to make records of all the developments and was doing so satisfactorily (Libyan Harvest Report, 1974).

Training programs

In 1975, Seedco sent an initial team of four farmers who (in conjunction with two demonstration farm staff) held a training school on the use of implements for shallow cultivation and combine seeding, and from those present selected a group of ten farmers whom they believed would best accept medic on their farms. The Seedco farmers moved from farm to farm helping each farmer to work his land and sow his medic and cereal. They reported '100% cooperation' from the Libyan farmers and they found them quick to learn how to use the new implements and eager to adopt the medic system. The combine seeders (sometimes called seed drills) worked satisfactorily

with farmers soon giving knowledge of adjustment of sowing rates and working depths. However the tendency was always to sow too deep and constant supervision was required for this ... Hydraulics [fitted to tractors to lift implements] caused considerable problems with poor quality rams and hoses and leaking fittings.

Farmers were 'most enthusiastic to use the Australian methods although use of heavy rates of fertiliser met some opposition'. The Australian farmer reporting on this year wrote

I made considerable efforts to get across the long term advantage of clover (medic) pasture on soil fertility and the resulting improvement in crop yield. The result of this will be in the second and successive years and I would therefore like to see some more assistance and supervision of these farmers next year. It would be a tremendous help if they were shown, as a group, the results on the Australian [the demonstration] farm, particularly to demonstrate what their own medic pasture should be like and how to graze it. Over-stocking could cause complete failure ... (Ianson, 1974).

They found that the Libyan planning for the year's program was much better than the first, when chaos had reigned at the wharves and the imported medic seed was only available on time due to the persistence of the Seedco farmers who manhandled the supplies themselves.

They welcomed the presence of separate machinery maintenance teams (sent in by the Australian manufacturers of the farm machinery being used) who were working with Libyan farmers to support them in their acquaintance with the new machinery.

The team in the following year were not so optimistic. The Seedco leader reported that the five Australian farmers arrived on time and that the seed was unloaded rapidly but took three frustrating weeks to clear through Customs.

The team leader in 1976 wrote that while the enthusiasm of a large number of the newly settled Libyan farmers towards adopting medic was not overwhelming, 'the shepherds have changed their opinions because they see the sheep doing well on medics'.

About twentytwo farms now had medic/cereal rotations. The Seedco leader wrote 'It is greatly encouraging to find some farmers who are capable of a very good job, some who are very keen and some who have detected the benefit that super and medics can bring'. However, he was disappointed that progress was so slow – so many farmers were still 'imprisoned' in the old system where the total area was treated with bagged nitrogen and returned low yielding wheat, and too many, in his opinion, continued to chisel plough 'as much as 250 mm. deep' despite the demonstration on the Australian farm. On the farms where shallow cultivation was used, germination occurred after only 20 mm of rain and good growth continued in spite of another four weeks of dry weather (Treasure, 1976).

In spite of this bout of disappointment, the Seedco farmers continued to work directly with the Libyan farmers – one farmer reports in 1979 working for five weeks helping to repair combine seed drills both in the workshop and directly on farms, helping to select thirtythree more farms for the introduction of medic pastures and to assist with the planting of medic and the application of fertiliser, of visiting fiftyeight farms at least once and on all farms trying to explain that cleaning drills was a necessity ('but without an interpreter it is very difficult') and of being sent to the Australian farm for four days to 'assess Polish farm machinery' being exhibited and tested. He wrote 'I enjoyed my association with the farmers, and found most of them eager to cooperate and accept my directions. I could have done a lot more for the farmers had I had a vehicle to use when I arrived, instead of being supplied with one when my term was more than half finished' (Schmidt, 1979).

Two colleagues were sent to work on a large barley and oat rangeland seeding project that was in progress. There they were required to supervise seven young tractor drivers. 'As they were quite competent at their task we spent most of our time sitting and watching them work. They used our tools to maintain their plant quite effectively'. This couple became disillusioned with their task and wrote that 'future seedgrowers teams would be better utilised if they were more closely associated with the farmers only' (Faint & Spencer, 1979).

Another Seedco farmer spent some time on twentyseven farms checking and repairing seed drills and scarifiers, then time on ninetysix farms planning with the individual farmers how to divide their land into medic/cereal areas.

Mostly the farmers were agreeable to our suggestions and cooperated, but in some cases we had to be flexible in our ideas and alter maps, but all farmers have agreed to sow 50% of their arable land to wheat and the remainder to medic, oats and barley ... Our visits to the farmers at this time were interspersed with calls for assistance with their repairs and we also advised every farmer who had not done so to commence working his ground for preparation for seeding particularly his medic area, which could be sown immediately. We completed all the farms with this service on the 4th of November, 1979. The next four weeks of my stay in Libya have been spent revisiting the farms and checking on seeding operations and crop establishment, with a few minor repairs being done as required.

He found it 'pleasing to return to these farmers and find them keen to show you their achievements ... I found the Libyan farmers very friendly and appreciative of our help and a pleasure to work with, although somewhat handicapped by my personal lack of communication with them' (Reichstein, 1979).

Fifteen Seedco farmers went to Libya in 1979. By this time a Seedco farmer (Lindon Richter) was employed by the Authority to be permanently in residence to monitor the progress of the project and liaise with the Authority. He reported that the allocation of Seedco farmers to the rangeland projects such as Benghazi Plain and Wadi Bab was not a happy one and that 'the men working in the El Marj area are obtaining much greater work satisfaction. Their efforts are appreciated by the farmers and there is much effective work that can be done, utilizing the capabilities of our men much better'. He noted that 'Spare parts have been the main problem encountered everywhere this year and to a lesser degree last year. The lack of parts and also the lack of cooperation from those concerned with their distribution has reduced our effectiveness this year' (Richter, 1979).

Three hundred Libyan farmers in the El Marj district had been helped in some way by 1979, and 130 had been given 'special attention' in managing their medic/cereal rotation. Richter himself made continued visits to these farms during the next three months to advise on 'weed control and medic management as well as seeder maintenance after seeding'.

The comparison between the Libyan profit gained from the increased productivity of the farms and the poor provision of funds for project resources called forth a comment from the sorely tried Richter.

The future success of seedgrowers teams depends largely on the Jabel el Akhdar Authority accepting the responsibility of providing the necessary vehicles and personnel/interpreters to fully utilise our expertise. The work is seasonal and must be done at the right time to gain maximum benefit. Most farmers could increase production by 100% if our techniques were fully accepted. The increased production of ten such farms would more than cover the cost of a 15 man team (Richter, 1979).

Machinery and problems with its operation

Gaining familiarity with new implements and the need to maintain them by carrying out simple, but prompt, procedures proved a central preoccupation of Libyan and Australian farmers during the length of the project. In spite of the greater care being shown to their machines by the El Marj farmers, there were 'latent troubles' with the combine seed drills. In the early days, few Libyan farmers cleaned the seed and super from the machine after use, and corroded distributors held tight by hardened fertiliser threatened immediate breakages when used again. The Seedco farmers set up a training school at El Marj to show the farmers how to clean and maintain their seed drills, and made follow-up visits to farms to inspect machines. A separate training program was made available to familiarise farmers using the new machinery for the first time. The training scheme comprised four sections covering tractor knowledge and care, tractor handling, drill knowledge and care, drill handling and cleaning. Twenty farmers attended and were given actual experience of each sector. The Seedco consultant, in an annual report, recommended that the Australian Farm be allocated a farmer-mechanic to 'constantly visit and assist farmers with their machines' (Treasure, 1976). Unfortunately the mechanics sent initially by the machinery exporting firms had not continued coming to Libya once the first demonstrations had been conducted and Bashir Joudeh had simply transferred their work to the Seedco farmers. The Seedco farmers resented this and some claimed that they felt they were not being utilised as farming experts to train fellow farmers in the new system, but were instead being used as machinery demonstrators and mechanics. They constantly recommended in their reports that action be taken to provide better means of organising spare parts and more efficient access to maintenance services.

The option of herbicides

While the technical team on the demonstration farm at El Marj carried out research programs to determine such matters as the optimum time of seeding on the demonstration farm, the Seedco farmers tended to worry about the need to train Libyan farmers in the use of herbicides to eliminate weeds, and often recommended that attention be paid to this. On their own farms in Australia they were entering a phase where the refinement of the medic/cereal system was being achieved through the chemical spraying out of grasses, in particular, to allow them to have almost pure medic stands. Yet attempts to introduce this more sophisticated phase of the system's evolution would only have put more pressure on Libyan farmers who needed at this stage to learn the elementary operations and management that could lift them out of the slough of an inappropriate and unproductive system. Fortunately, perhaps, the Seedco farmers were too occupied with their immediate formidable task to push the idea further.

Shearing displays

It was a similar degree of blindness to the fundamental objective that led the Libyan authority, in the same year, to import two Australian shearing instructors to El Marj to train Libyan shearers to shear sheep, but with little success. Most of the recruits were itinerants who moved on and were not available either to train others or to work the following year, and 'Generally speaking there was not much enthusiasm amongst shearers as I feel they were under paid and there was no incentive to work hard – the pay was the same as other less demanding tasks' (Salter, 1976).

Benghazi Plain rangeland development project

In 1975 the Seedco team sowed 6000 ha of barley undersown with 7 kg/ha of medic and 50 kg/ha superphosphate. They mixed the medic (cultivars Jemalong, Snail and Harbinger) with the superphosphate and sowed it through the seedboxes of wide-line combine seeders (Schultz, 1976).

The Seedco farmers persuaded the Libyan director not to plant cereals in strict rotation with the medic pasture because they believed there was too little rain to make cereal production profitable. The director was persuaded and they sowed more medic in conjunction with oats for hay and applied a dressing of superphosphate. They sowed and dressed 2000 ha. The director had assumed from his reading of the 'ley farming' system that wheat must follow medic.

The Benghazi Plain medic planting went well and provided an estimated yield of 400–500 kg/ha of medic pods. The cultivars Snail (*M. scutellata*), Harbinger (*M. littoralis*) and Tornafield (*M. tornata*) were planted on a further 12 000 ha. In the fourth consecutive year of this particular rangeland project 'the depreciation to machines, especially drills in this area is enormous, caused by travelling at high speed over very stony ground and contrasts greatly with the care being increasingly shown by the farmers at El Marj who work their own machines'. In spite of the damage to machinery the 'results of past years work are being seen in areas becoming more arable, increasing yields and high numbers of medic pods in many areas'.

The Seedco farmers were asked to sow 6000 ha of annual medic at Wadi El Bab, another rangeland area, which they did successfully. The Libyans were planning on turning part of this rangeland project into a medic seed producing project and placed orders for two suction seed harvesters from Australia. In 1979 a further 4500 ha were planted to barley and medic on the Benghazi Plain.

No report is available about who exploited the pasture on the Benghazi Plain or how much, if any, grazing took place. It appears from what evidence is available that large quantities of hay were made as yields of both medic and cereals were good, yet there are also reports of good regeneration of the pasture during the period under review.

At Wadi Kharoubeh, where medic was planted on the flood plain where it received natural irrigation several times a year, there was a successful integration of the existing livestock holders into the program of establishing and exploiting it. This was carried out by a Libyan director who worked directly with the local communities to conceive a division of land and pasture and to institute a set of penalties (mainly based on withdrawal of animal health services) for pasture poaching and/or fence cutting.

Seedco farmers and the demonstration farm

By 1979 the General Manager of Seedco (who visited each year) was becoming very critical of the entire demonstration farm program. He wrote that

If Bashir Joudeh . . . wants the [demonstration] farm as a large scale production unit and show place for visiting dignitaries, then all is well. If on the other hand the Libyan farmers are to benefit from the Ley method of farming, the use of Annual Medics and Fertilisers, then there should be much more communication with the Libyan farmers by personnel of the South Australian team . . . Field days on the Australian farm should be a feature; at present this is not so (Farnan, personal communication to Minister of Agriculture, 1979).

The demonstration farm staff made demands on the Seedco team that were resented.

Due to lack of experienced personnel on the Australian Demonstration Farm two men have been involved there for the last five weeks doing the actual seeding. It is unfortunate that these men were not available for work among the farmers. If men come in the future I would recommend that more be stationed in the El Marj area and to only be associated with a few project areas (Richter, 1979).

Another Seedco employee wrote that

The Australian agronomist [on the demonstration farm] should be more forceful in implanting these [methods of using medic] to his counterpart . . . this man [the counterpart] obtained his knowledge at a University in Libya, mainly from American text books which do not apply under their climate or conditions.

The Seedco management in Adelaide offered to cooperate with the South Australian Department of Agriculture in arranging a study tour to Australia for the Libyan counterpart.

The writer also believed that the research work on the farm received too much attention – 'trial plots only proved departmental theories' and he reflected ex-farm manager Bill Kelly's worry that 'The contact of the Farm team with the Libyan farmer is negligible'. The Seedco farmer compared this arm's length relationship maintained by the demonstration farm team with the constant, daily contact of the permanent Seedco representative (Richter) who was known by all and, in turn, had become privy to a lot of knowledge about the local farming. (Seedco report, unsigned, 1979).

More pressure on Seedco farmers

By 1979 the Seedco farmers were finding their task was outgrowing their resources. The Seedco farmers still came in in small teams for the sowing of medic and cereals, and Seedco farmer Lindon Richter was living all year round at El Marj. The difficulties in getting spare parts for the machinery (damaged either owing to inefficient cleaning or driving too fast) were causing frustrating delays, mainly due to individual Libyan officials who would withhold parts for no apparent good reason, and this led to many irritable comments. But it was the repairing and setting up of equipment that took up time and really frustrated the Australian expert farmers who should have been spending more time evolving management regimes on the farms.

One farmer (who later carried out two intensive training schedules in Morocco) wrote 'I was under the impression that Australians were required for technical assistance, but now realise that my job is mainly as a field mechanic'. He was sent to Wadi Karoubeh to oversee the planting of the large wadi bed with medic.

With 9 tractors and machines in the field it is a continuous job keeping all wheels rolling. Even the simplest repair jobs require my assistance as none of the men are provided with tools ... I always supervise and assist with filling the spreader to ensure that the medic is mixed thoroughly and occasionally drive the spreader while the driver has his meals (Masters, 1979).

Farmer training after 1980

After the demonstration farm team were pulled out by the South Australian Government in 1980, Seedco were asked by the Jabel el Akhdar Authority to take over the management of the farm and this added to their responsibility.

Requests by Seedco for mechanics with farm machinery expertise to take over the role of setting up of seeders and repairing machinery were ignored, and the Seedco farmers continued to be frustrated because they spent more time repairing machinery and setting up drills than working with the Libyan farmers to develop a functional farming system based on medic.

However, a lot had been achieved. In 1981, Seedco farmers reported that 'very few farmers were cultivating too deeply', 30% of the farmers had their machines in 'excellent order' prior to seeding on 10 November, approximately 40% of farmers in one sector were interested in medic and eight farmers (in one sector alone) had planted an additional area of pure medic – as distinct from the majority who tended to still plant it with barley and/or oats. In another sector five out of fifteen farmers sowed pure medic and about 98% of farmers in this sector used fertiliser. In another sector many farmers sowed medic and oats for cutting and baling but 'Others are accepting the Australian cropping system and setting aside an area each year for growing (medic) and the following year this area will be sown to wheat or barley. Most farmers do sow fertilizer with the grain'. In another area there were seven farmers who used medic and 'were well aware of its advantages' and another twentynine who were planting some for the first time.

In 1981–2 alone, over 600 farms had been visited by the Seedco team, machines repaired and prepared for seeding, and advice given to Libyan farmers who were growing medic. Many of the combine seeders were in good and clean condition; about 30% required new spare parts. Most of the farms were using shallow cultivation as recommended, although some were still cultivating too deep.

The Seedco farmers regretted not having more time to help their Libyan counterparts to refine the management of their medic/cereal systems. Reporting to the Jabel el Akhdar Authority on the state of the project at the end of 1982, a Seedco farmer consultant commented that the yields of cereals and medic on farms continued to be consistently high and certainly much in excess of crops planted using the existing method. 'Generally practical skills of soil preparation, weed control, and time of sowing have resulted in crops now appearing vigorous and healthy, and the value of the Medic–Cereal rotation is evident in the excellent crops on the farm'. Some Jemalong (*M. truncatula*) had been attacked by a medic grub, but other cultivars such as Paraponto (*M. rugosa*) and Snail (*M. scutellata*) had proven to be resistant and were untouched.

The consultant noted that the major part of the demonstration farm was to be allocated to local farmers but he suggested that two farm sections from it should be retained as a demonstration base. He also recommended that sheep be reintroduced to the demonstration farm and a reliable shepherd be employed to care for them. Seedco were worried that the 'optimal benefits' of the excellent pastures were not being obtained.

This requires good management of grazing pressure to hold pastures to approx. 5 to 10 cm height, resulting in reducing the presence and seed set of weeds and increasing the benefits from the medics. The divorcing of the sheep from the (demonstration) farm has prevented this being possible, although visitors may be taken to the demonstration unit at 223 Selina where near perfect results are being obtained. [The pastures] are thriving and contain an excellent proportion of medics. Weed control in pastures ... is very effective (Treasure, 1982a).

In his report to Seedco on the future of the farm the consultant recommended that if a flock of local sheep was re-introduced to the farm it should be improved by selection and breeding.

In addition, the medic/cereal rotation should be continued on the remaining demonstration farm base but there should be an area put aside for the demonstration and an assessment of alternative systems (presumably the continuation of the control of wheat/wheat) and there should also be some investigation of forage crops such as faba beans and more effective varieties of wheat and insect resistance medic.

The condition of the machinery was reported to be 'much improved and the deteriorated situation that greeted the first team of experts has been somewhat rectified' but 'difficulty still exists because of non-availability of parts'. Seedco wanted to see some trials set up using 'weedicides popular in South Australia' but most of all hoped 'that methods enabling the farmers to benefit from knowledge that is available to them be improved' (Treasure, 1982b).

A review of project results

The immediate availability of Australian designed and made scarifiers and combine seeders to the individual Arab farmers overcame most of the resistance to the abandonment of deep ploughing, but it was the Seedco farmers working side by side with the Arab farmers who inculcated the knowledge of how to operate and care for the new machinery. Without this face-to-face demonstration and training component it is probable that the Arab farmers would have removed the depth wheels on the scarifiers and used them to plough deeply as they did elsewhere, and that the combine seeders provided would have quickly become broken and been left in sheds unused, as later occurred in Algeria.

The establishment of the medic pastures was accomplished in a satisfactory manner. In the Jabel el Akhdar, it was the farmer himself who prepared the land and sowed the pasture and then the cereal with the advice and assistance of the Seedco farmer.

The decision of Bashir Joudeh to use the Seedco farmers as his extension and training component was vindicated. The problem was that, in refusing to continue to employ mechanics for the continuing machinery and equipment service, the Authority frustrated the Seedco farmers' intention to move into pasture management work with the Libyan farmers, and this left the Arab farmers floundering on the livestock side when the Seedco farmers were withdrawn. This does not invalidate the decision to meet the problem of an almost non-existent institutionalised extension and advisory service by engaging expert farmers to work with Arab farmers. The direct communication of common sense and experience was much more effective than a hesitant intervention by an inadequately trained extension agent.

There is no guarantee that the existence of an extension department will result in a continuation of assistance to farmers especially when new management regimes are being introduced and adopted. As all project staff know, the constant turnover of extension staff in developing countries means that farmers can seldom rely on a continuity of stable and relevant advice.

What the Seedco expert medic farmers did provide was a model of farmer training that, in conjunction with the provision of appropriate resources, proved capable of bringing about a radical change in farm management and leaving behind a community of farmers confident enough in their new system to continue with it once the foreign experts left.

Expert farmers vs. technical staff

The expert farmers brought in to train Libyan farmers in El Marj reasoned that the farm was supposed to be an actual demonstration of the farming operations and management required for a medic/cereal system and they were convinced their own expertise was more relevant than the technical expertise of those engaged to carry out research programs. It was after all the Libyan Government's objective to change the existing farming system and introduce the new one so as to rapidly increase overall productivity and this could best be achieved by showing the Arab farmers how to perform the basic farming operations of the new system.

Such change would not, in the first place, come from small plot experiments and selection programs that were designed to extend the productivity of the system on the margin (Personal communications from farmers involved).

When the official Australian interest in the projects waned, Australian farmers were left to carry on. In the case of the El Marj demonstration farm, it was the Seedco farmers who looked after it when the technical experts returned to Australia, combining this work with their own extension program, and they were left to pack it up in 1984. In the case of the Gefara Plains project, all the technical advisers returned to the Department of Agriculture in 1980 and it was an Australian farmer (Geoff Collins) who continued the work of expanding the project and recording the results until he was recalled in 1983.

The Australian farmers were quickly convinced of the suitability of the system to Libyan conditions and the results obtained during the ten years they worked with Libyan farmers supported their conviction. Australian farmers saw their Arab counterparts as a dynamic and responsive force, whereas the technical teams saw them only as arm's length passive receivers of expert wisdom.

The Australian farmers were, at times, exasperated beyond measure by the frustrations they experienced. Such frustrations, however, did not undermine their firm belief, drawn from their own day to day experience of working with the Libyan farmer, that the Libyan farmers were capable of operating the system and that, given half a chance, they could turn around Libya's poor agricultural performance.

Conclusion

Essentially the Libyan projects had demonstrated:

- That the cereal phase of the medic rotation could be successfully used by Arab farmers if they had scarifiers and, to a lesser extent, combine seeders, and were shown how to use and maintain them;
- that the grazing phase using local sheep could bring about a great improvement in nutrition and production;
- that farmer training using expert medic farmers was quick and effective in achieving results;
- that rangeland pasture production could be achieved quickly and cheaply using medic and its associated sowing techniques.

In 1987, the Syrian based Arab Centre for Studies of Arid Zones and Drylands (ACSAD) funded a study of the Libyan medic/cereal projects. After visiting the farms, the authors concluded that the economic advantages of a medic/cereal rotation, even in the manipulated agro-economic context of present day Libya, in which subsidies heavily favour the production of cereals, remain superior to any other system, but that the absence of supplies of medic seed and the lack of any support services specific to a medic/cereal system from within the Ministry of Agriculture were causing farmers difficulties and inhibiting the expansion to other farmers of the knowledge and experience gained (El Akhrass, Wardeh & Sbetah, 1988).

Projects elsewhere in the region did not make scarifiers and combine seeders available to farmers and the pasture phase often failed due to rough and cloddy tillage.

Although the commercially available Harbinger cultivar provided everything and more required of it on the Adjulyat rangeland site and no difficulties were experienced in the cereal zone with other cultivars imported from Australia, Australian scientists and technical experts began to talk of the problem of 'adaptability' to North Africa of medics produced in commercial quantities in Australia, forgetting that this was their natural habitat, and set an unfortunate emphasis on the selection of local and specific cultivars as a necessary precursor to the successful establishment of the medic system on farms in the region.

In spite of the demonstrated success of the Seedco project in training the Libyan farmers to become familiar with techniques and farm operations necessary to the new system, there was no attempt on future projects under the guidance of Australian technical experts to undertake a similar program. Expert farmers were employed to drive tractors but not to advise or assist other farmers.

6

A medic project in Algeria

Introduction

While the Libyans were establishing their varied and somewhat radical programs of farm improvement using medic pastures, the Algerian Government turned its attention to introducing medic too, but in a more conventional manner.

Background

The Libyan projects were inspired by observation of the medic system in operation in Australia. It is not clear what sparked off the FAO/Algerian interest in the system. It may have originated from a small project in Tunisia undertaken between 1971 and 1974 by Dr John Doolette, a South Australian agronomist, who had been on the staff of the Turretfield Research farm, and who went to work for CIMYTT.

The joint Tunisian/CIMYTT project was to increase cereal yields by replacing fallow with medic pastures on some cooperative and private farms in the El Kef district of Northern Tunisia. Dr Doolette worked alone, training Tunisian counterparts as he went along. The ordinary commercial varieties of medic, Harbinger (*M. littoralis*) and Jemalong (*M. truncatula*), imported from Australia established well on all sites.

The profitable exploitation of the medic was not so straightforward. Dr Doolette found many barriers in the way of the farmer who wanted to increase his productivity and reduce the erosion on his farm by using medic pasture. Deep ploughing created a rough seed bed that was not helpful to the germination of medic. When used for cereals it prevented good regeneration. However, Dr Doolette saw no reason why the implements being used on farms could not be adjusted to carry out shallow cultivation. He found worrying the 'attitude' of many of the Ministry officials who could not understand the 'logic of the system' and who failed to organise the distribution of farm resources efficiently.

Dr Doolette was employed to introduce the medic as a source of green manure for the cereal crop and he had no quarrel with that point of view.

In North Africa ... [he wrote] ... the rotation must be focused in its conception toward the cereal crop. This is regardless of whether the farmer is more or less interested in animal production. He may easily emphasize the animal production or the crop production, whichever he prefers. But in conceiving the rotation he [the farmer] must focus on the cereal crop. If the focus is on the cereal crop, the gain in animal production will follow automatically. If the emphasis is weighted too heavily toward animal production, cereal culture will invariably suffer and decline. This would probably represent a misallocation of resources within one of the country's major cereal zones (Doolette, 1980, p. 77).

This early project left behind a pocket of interest within the institutions of Tunisia.

The project raised interest in neighbouring countries because politicians were under pressure to increase yields of cereals and meat production. Officials from Ministries of Agriculture throughout the Maghreb and the Near East were sent to visit Australia to discuss technical cooperation or negotiated with international agencies to establish their own projects in which medic was a component.

The Algerian Government applied for and received support from the Food and Agricultural Organisation of the United Nations to set up their project which began in 1973 with the objective of increasing cereal yields by introducing medic and sub-clover pastures as a green manure in place of fallow.

Post-Independence Algerian farming

Wheat yields in Algeria in 1971 were on average 614 kg/ha, little changed for forty years, and there were just over ten million sheep being kept on range and farmland. Sheep were fed poor quality straw and grazed sparse weedy fallows in the cereal zone in the summer, and even sparser vegetation in the rangeland in winter and spring, supplemented with grain. Their best quality local feed was the cereal stubble, which had large quantities of grain in it due to inefficient harvesting. Mutton is traditionally one of the staples of the Algerian diet so the poor level of production was of great concern to the government.

Algerian farming is closely supervised and influenced by the Ministry of Agriculture and Agrarian Reform (MARA) and its associated Institutions. Land reform, designed to return land to the farm workers after Independence from French rule, has been somewhat haphazard. Some pre-Independence French farms and estates were converted to State farms where the workers were persuaded to give the farm they had worked on to the State, and some to Cooperatives (Autogères) where the workers formed groups to retain the land and work it themselves. In the poorer farming areas there were many Algerian farmers who retained their smallholdings unchanged by the revolution that accompanied Independence. Whatever the size of the State or Cooperative farm (and some of them were very large – 2000 ha and more) their management was directed from Algiers, either by regular directions to the Farm Manager or, as in the case of the Cooperative, through the appointment of a Ministry agronomist who 'advised' the Chairman of the Cooperative (rather like the British Political Officer in the Princely State in India). The Ministry decided a policy of desirable production targets for particular crops each year and the administration to achieve them was the responsibility of the regional and district officials.

The Ministry supplied cereal seed, fertiliser, and concentrate for animal feed and directed how much each animal should have each day, allocated farm equipment and kept the cheque book. The farm program for each season was monitored by officials from the local agricultural office, as well as from Institutes for cereal production (Institut Développement des Grandes Cultures), sheep production (Institut de Ovine) and cows (Institut de Bovine). Farm machinery came from the Office National des Machines Agricoles (ONAMA) and cereal seed was produced and distributed by the Office des Cèrèales. Technical advice comes from the Research Centres of the Institut de développement des Grandes Cultures (IDGC – that later became the Institut Téchnique de Grandes Cultures, ITGC).

The State and Cooperative farms were lavishly equipped with machinery and implements and with labour and followed the traditional French farming system of cereal monoculture based on fallow and deep ploughing. On State farms of 400–500 ha, we frequently counted eight large tractors and a number of smaller tractors and four mechanical harvesters as well as quantities of deep ploughs, chisel and disc ploughs, harrows, and seeders as standard equipment. This is four times the level of mechanisation one would expect to see on comparable farms in Australia.

Animals, if kept, consisted of a herd of dairy cows made up of imported breeds kept and fed in sheds, and a small flock of sheep (about 200) that were fed on concentrates and baled straw and grazed the sparse fallow and cereal stubble. The number of families involved in the Cooperative could be as many as thirty and the work force on the State farms was often between forty and fifty workers. In addition, many of the cereal farms were opened to nomadic flocks from the rangeland after the cereal stubble had been harvested. If this access was permitted, it was organised by Ministry officials. Nomads had to pay a fee to graze and were allocated fields by the official responsible.

Following Independence from the French, the Algerians continued the uneasy reliance on a clumsy, costly and environmentally damaging farming system to produce cereals and environmentally damaging seasonal transhumance of politically alienated nomads and their livestock for the bulk of their sheep meat. In addition, however, they socialised the agricultural sector and in the process established a technocratic nightmare that continues to inhibit production even today.

The project plan

The objective of the FAO project in 1973 was to replace the long, cultivated bare fallow common in the cereal zone of Algeria with a medic pasture phase.

This project was planned in the conventional manner so familiar to those who work for large development agencies. An identification mission was carried out using a miscellaneous and anonymous group of consultants. They recommended that the project team should work directly with the managers of State and Cooperative farms. Tight management control of farming operations of the type then operating in Algeria was considered by development agencies to be an advantage when introducing new techniques. It was believed that the private farms round about would copy the State/Cooperative farm if the system was successful.

A list of expertise required for the project was defined (all experts had to be recruited by and acceptable to FAO) and a machinery list drawn up and orders placed. The project was named 'Development of winter cereals and the replacement of the fallow' ('Développement des Céréales d'Hiver et Suppression de la Jachère, MARA (FAO) ALG (71)537 duration 1973–1977') and it was to cover 30 000 ha (Bakhtri, 1980, p. 123).

Sites were selected in the districts of Oran, Algiers, Constantine and Annaba. They are considered to be representative of the four major cereal growing areas in Algeria with average annual rainfall of between 350–500 mm and altitudes up to 750 m (Webber, 1975, p. 3).

Selection of project team

The FAO/MARA project had funds to employ four technical experts (three agronomists and an agricultural machinery specialist), a bilingual administrative secretary, two or three special consultants, and to buy equipment and machinery with which to carry out the field work on the designated domaines and State farms. Three expert agronomists were to train five Algerian extension agents (three of whom were based in Constantine, Oran and Annaba and two in Algiers) in the establishment and management of medic pastures. Only two agronomists were eventually employed and the other position was used to employ a livestock expert who had experience in Australia.

The machinery expert was to work in each of the centres to demonstrate and train operators in the use of the machinery supplied and to report on the state of Algerian agricultural machinery supplies and performance generally. No expert dryland farmers from Australia were employed. It is not known if any of the permanent staff had special experience in compiling extension programs (DCSJ, n.d., p. 10).

Australian involvement

In August 1974 the Secretary General of Agriculture in Algeria visited South Australia to ask for cooperation on a 'development basis'. In particular, Algeria would like expert farming advice from South Australia – in fact, he had a list. He would like a soil and water conservation expert, a sheep production specialist, and one or two 'practical farmers' from South Australia to go to Algeria to advise Algerian farmers, 'similar to Libya' (Preliminary Proposals, 1976).

By this time the first results being obtained from medic pastures sown by the Seedco farmers in Libya in 1974 were causing comment throughout the region, and South Australia was playing host to a succession of high ranking Ministry officials representing Ministers (who came later) intent on finding out if Australia had an answer to their farming problems. They were prepared to negotiate terms and initiate trade.

This new level of interest took the desultory exchange of information out of the inner recesses of the Departments of Agriculture and it acquired a prominence that attracted the interest of the more dynamic units of government (in particular the Departments of Trade and Development) and the commercial community. They were interested in the opportunity a transfer of 'dryland farming expertise' might offer to trade.

A fairly broad (informal) agreement on technical cooperation with Algeria was initiated with the Premier's Department in South Australia and two South Australian consultants (agronomist G. Webber from the Department of Agriculture and Dr E.D. Carter, a lecturer in pasture agronomy at the Waite Institute of the University of Adelaide) were employed as consultants to the project.

It has been possible to recover the final reports from the agronomist, the livestock expert and the machinery expert and one of the two Australian consultants employed to give an overview of the project's operations and to recommend the future directions for a next phase.

The medic pastures

The first step was to sow the medic and sub-clover to see if it would grow.

The soil types on the farm sites were identified as suitable for the Australian medic cultivars Jemalong (*M. truncatula*), Harbinger (*M. littoralis*) and Snail (*M. scutellata*). One sub-clover Clare (*T. brachycalcinum*) was also included – a sub-clover that is commonly used in Australia with medic cultivars in regions of relatively high rainfall (e.g. 450 mm) (Webber, 1975, p. 5). In the 1974/5 season 1700 ha were sown to pasture – 1610 ha to medic and 90 ha to sub-clover (Webber, 1975, p. 10).

The cultivars from Australia performed well on all sites in the establishment year in spite of a late seasonal opening and below average rainfall (Webber, 1975, p. 3).

(a) Medics and the high plateau

The team leader, who was an agronomist, had some reservations about the suitability to Algerian farms of the cultivars Jemalong (*M. truncatula*) and Harbinger (*M. littoralis*). He agreed that they performed well in the warmer regions where winters were not excessively cold, but he thought they may be at risk from frost in altitudes above 600 m. He recommended that for the cold regions of the High Plateau of Algeria (where considerable cereal production takes place, e.g. Tiaret in the west and Setif and M'Sila in the east) it would be necessary to search for good local ecotypes. Later research and experience proved this to be an opinion without foundation as a number of local ecotypes were found flourishing freely above 600 m, well able to survive very cold winters, but research and broad acre sowing also confirmed that Jemalong (*M. truncatula*) performed very well on sites in Tiaret (Adem, 1989, p. 9). Unfortunately the doubts expressed in the early report were often used to support a widespread belief in the region that medic would not grow above 600 m.

(b) Local medics

The Algerian project proved to be a forerunner of the continuing debate as to whether specific cultivars should be sought for specific sites through programs of research, identification, selection and multiplication before widespread use was advocated, or whether it is more practical to encourage what already grows *in situ*, initially increasing the seed bank and raising soil fertility by sowing a commonly available commercial variety and topdressing the site with superphosphate.

The existence of local ecotypes on farms involved in the project was noted by the consultant agronomist, G. Webber, during his visit to the sites. He observed that 'sufficient natural Medicago species occur in many fields to develop into adequate pasture', and suggested that 'If some fields could be selected, topdressed with superphosphate and managed for maximum seed set, considerable information could be gained regarding the possibility of rapidly expanding the programme with these spontaneous species'. He added the reminder that 'If followed by shallower cultivation techniques and alternate year cropping the full potential of these fields could be realised' (Webber, 1975, p. 8).

Later, in 1979, on an inspection of one of the original project Domaines at Khemis Miliana (Domaine Chouhada), we observed the dominance of local cultivars (about 75% of plants) in fields originally planted to Jemalong, which is easily identified by a distinctive leaf marking. On later visits in 1981 and 1983 there was little sign of Jemalong but the local varieties were providing abundant pasture to keep a flock of two thousand sheep over the entire year and sufficient nitrogen for the subsequent cereal crop sown on the farm (notes of Chatterton & Chatterton, 1979, 1981 and 1983).

The selection of inappropriate machinery

The cereal phase got into trouble from the beginning owing to the absence of scarifiers for shallow cultivation and combine seeders for shallow seed bed preparation. Those who ordered the machinery for the project were unaware that implements designed for deep ploughing were different to those needed for shallow cultivation.

The machinery provided for the project was trenchantly criticised in the final report by the British machinery expert.

Twelve expensive pieces of equipment were supplied. There was a disc-seeder so big that it was too large for easy use and so designed that it was extremely difficult to empty and clean, and the machinery expert was forced to modify the machine himself to enable it to be cleaned satisfactorily. The trailer was supplied without a braking system. Not only was that illegal in Algeria but it was also very dangerous in the mountainous terrain in which it was to be used. The disc ploughs ploughed too deep for the medic to regenerate, the seeders were too slow, too lightly built and easily broken. A fertiliser distributor had a feeble frame that made it extremely susceptible to fractures caused by bounding over the deep furrows left in the soil by the disc ploughs. The deep furrows also had an effect on the distribution of seed from the disc seeder so that potential yields were reduced due to the maldistribution of the seed (Muckle, 1978, pp. 4–6). The tractors supplied were the only ones of their type in Algeria and their hydraulic system needed a special type of oil that was not available in the country, and they deteriorated quickly. The harvester was designed for heavy European crops harvested with a grain humidity of 15–16%. In Algeria where harvesting of relatively light crops took place in much hotter summers the humidity was seldom above 11%, the machine broke about 50% of the harvested grain. In addition, the machinery purchased did not conform to the metric system which Algeria has adopted and this led to many problems with measurement and calibration.

There were also problems with the supply of spare parts, (Muckle, 1978, pp. 25–6). The uneven soil, the result of deep ploughing, caused problems with the harvester and Muckle observed quite large losses of grain that were confirmed after the following autumn rains when the regrowth of cereal from this lost grain was significant (Muckle, 1978, p. 6).

The equipment purchased for the project was clearly unsatisfactory. The machinery expert was later sent to Australia to inspect equipment manufactured specifically for a medic/cereal rotation and he presented a list of appropriate equipment and an analysis of why it was needed as part of his final report. The report was printed in Rome in 1978 but the funds for equipment had been spent in 1973 and unless a second phase of the project was to be funded there was no possibility of the relevant purchases being made (Muckle, 1978, Annexe 3, pp. 28–36).

Deep ploughing vs. shallow cultivation

The machinery expert (in conjunction with others involved in the project) carried out a series of trials that indicated that different depths of cultivation made little difference to yields. The type of implement used in these trials is not specified in the published results. A comparison was made of the time taken to cover a specified area of ground and this showed a large advantage in favour of shallow cultivation, but there is no data relating to the comparative costs incurred.

Without scarifiers available for the cereal phase, the medic failed to regenerate adequately after the initial years due to the deep ploughing which the farm managers insisted on carrying out (Webber, 1975, pp. 3, 5).

The practice of deep ploughing is constantly reinforced in Algeria by the French textbooks used as the basis for the education of techniques (DAWPRW, 1979, p. 111). Extension material from the ITGC continues to this day to say that deep ploughing is necessary to enable the cereal root to seek deep moisture and develop a root system prolific enough to sustain the plant in good health (ITGC, 1987).

The project team were unable, given the resources available to them, to successfully establish a regenerating pasture except on small areas where they themselves controlled the conduct of the plot.

Only on one cooperative farm did the system become established due to the perseverance of an Australian-educated Algerian agronomist, a progressive Algerian farmer, and the lucky acquisition of an Australian scarifier. On this farm (Domaine Chouhada, near Khemis Miliana) a medic/cereal rotation was established and its success as a replacement for the traditional bare fallow demonstrated for almost a decade. The original success was probably due to the fact that they were able to obtain a scarifier designed for shallow cultivation.

In 1978 an Australian firm, John Shearer Pty. Ltd, sent in some equipment – a combine seeder, scarifier and harrow – that became subsequently the focus of a quarrel over whether it was a gift (in advance of substantial orders) or a sale. The farmer/manager of Domaine Chouhada used the scarifier for a number of years to prepare a shallow seedbed for the cereal phase of his 250 ha of cereal/medic rotation. The Australian combine seeder and harrow set were later seen being used with good effect on the ITGC Centre du Récherche at Constantine to sow several hundred hectares of good medic pastures.

There was no evidence of overgrazing of the medic pods on the farm sites that could have accounted for poor regeneration – in fact, the very opposite. The lack of sufficient sheep to graze the pastures caused livestock experts to make a point of asking that in future projects medic pastures should only be sown on farms where there were sufficient sheep to graze them properly (Pattison, 1978, p. 72).

Weed control after medic pasture

The machinery expert was unequivocal in his belief that the supply and use of inappropriate machinery was the major reason for the widely held opinion current at the end of the project that lower wheat yields and increased weed infestation were the result of the introduction of medic pastures.

The agronomist agreed that there was no justification for the opinion that medic brought in weeds refuting it with the claim that the weeds were already there before the medic was sown and that it was ineffective cultivation of the soil and inadequate grazing that led to weed infestation. He suggested that the introduction of a chemical herbicide program and a mechanical topping of weeds during the pasture phase would be useful ways of dealing with the problem until the appropriate equipment was available for efficient cultivation (Golusic, 1978, p. 59). This was at most a short term evasion of the fundamental problem but without the scarifier to carry out the proper operation what could the team do?

Livestock and medic

Although the objective of the project was to increase cereal yields this did not take place due to the inability of the team to carry out shallow cultivation and thus control weeds. However, the medic pasture did provide an unexpected success in feeding livestock. 'In spite of the short duration of the project (three seasons - 1973/4, 74/75 and 1975/76) ... On all the farms where medic was established for livestock ... thanks to the culture of medicago the keeping of sheep had become profitable for the first time in 1975/6' (Golusic, 1978, p. 58).

Just as the Western Australian team, sent in to increase cereal yields using medic pasture as a green manure in Libya, found it provided a startling increase in actual and potential livestock production, so the FAO team found it in Algeria.

When in 1977 the project's livestock expert wrote his final report, he argued that a cereal/livestock system based on medic would make available for the first time in the region a truly integrated (that is, non-exploitative) system of sheep production rather than the conventional competitive (and therefore exploitative) system.

In his opinion 'the growing plant of the annual Medicago species is naturally the centre piece of the whole crop, medic rotation system' (Pattison, 1978, p. 8), and he proved his point in an articulate and painstaking exposition shown below of the performance of medic on the domaines involved in the project.

He wrote that

In the farming sense in Algeria the cereal–livestock system based on medic pastures would be an integrated system. The medic pasture produced benefits for both the crop and livestock:

for the crop, soil fertility and soil structure is improved;

for the animal, there is an increase in the quantity and quality of the feed per head and also an improvement in the time of availability of the feed.

The integration is carried further because grazing assists in the recirculation of nutrients adding to the benefits obtained by the crop. Cropping at intervals stimulates pasture growth which would otherwise decline over a period of years because of the lowered vigour of the medic as soil fertility increased.

Because of the increase in soil fertility the system is a stable one.

It may be possible to develop other livestock–cropping systems which are capable of supporting as many animals as the medic crop rotation but in which the feed is produced by annual fodder crops – for example oats, barley, berseem clover, etc. Such a system would not be a truly integrated system. Rather it would be a cropping system in which a large proportion of the crop grown each year is fed to livestock. Livestock would compete against cropping in that:

(i) Relative profitability of selling crop products for cash rather than feeding them to stock would be constantly under review.

(ii) Soil preparation and sowing of forage crops would be competitive for time and resources with the sowing of cash crops.

He pointed out the exploitative nature of growing forage crops in the dryland conditions common to the cereal zone of Algeria.

A system of increased cropping intensity geared to producing feed for livestock would be an exploitative one. Increased quantities of nitrogen fertiliser would be needed to ensure satisfactory crop growth. Also soil structure would be further damaged by the increased cropping intensity. Certainly the growth of large areas of fodder crops in place of fallow would answer the criteria of FAO Project ALG 71/537 'Developpement des Céréales et suppression de la Jachère'. However it would not fulfil the criteria of the proposed FAO project ALG/77/028 'Integration of Cereals and Livestock'. Rather it would introduce a system which is unlikely to be a stable one because of its exploitative nature and one which is centred on competitive enterprises trying to exist side by side (Pattison, 1978, p. 2).

This logic remains unrecognised by the agencies still trying to increase livestock production through the use of sown fodders in the region.

Yield from medic pastures

Arguing that it is not only the yield of the pasture that is important but the time at which it is available for the livestock, Pattison explained the way in which both the dry and green matter produced by the medic pasture could be exploited in the existing conditions.

For example, he demonstrated that in the 400 mm rainfall zone it was not unreasonable to expect 10 tonnes/ha of dry matter from medic if rain fell at the most favourable times. In practice (and this is the reflection of the capricious rainfall pattern of most semi-arid regions) the yield is most commonly between 4 and 7 tonnes/ha. The comparisons of yields from medic, grasses and broadleafed weeds common to fallow on nine of the Domaines in the project in 1974–5 revealed that the medic provided 10.7 tonnes/ha of green matter and 2.68 tonnes/ha of dry matter (estimated); the grasses provided 7 tonnes/ha of green and 1.75 tonnes/ha of dry; and the weeds 1.6 tonnes/ha of green and 0.4 tonnes/ha of dry. In the Oran region in the same year, the green yield of medic was up to 35.4 tonnes/ha of green matter while the weedy fallow produced barely half of that.

In the El Khemis area, a field on Domaine Chouhada was evaluated. After an original planting of Jemalong (*M. truncatula*) the pasture had been left for two years without the intervention of a cereal phase and had become dominated by a local ecotype of *M. polymorpha*. Even after a considerable period of grazing (the medic had been available to the livestock from December on) the yield of dry matter was 4.8 tonnes/ha when it was cut on 3 May – six months later.

In a grazing trial on another domaine on which there was 317 mm of rainfall during the growing season in October to mid-May, the carefully controlled production figures were 2.46 tonnes/ha of medic and 2.42 tonnes of grass species (mainly barley) (Pattison, 1978, p. 10).

Seasonal availability of medic

It is the provision of good quality forage at appropriate seasons that make medic pastures so valuable for a dryland production unit. While Pattison is careful to say that 'the vagaries of the weather ... mean that shortages of pasture can occur at any time of the year' and that spring droughts have the most potential for serious problems, he affirms that medic has the growth pattern and fundamental characteristics to provide the most favourable pattern of supply of all forage plants.

Given a January–February lambing time, medic pasture should match demands from February to July inclusive without any other source of feed being needed, either as green growing feed or dry residue (Pattison, 1978, p. 22).

From August to October the dry medic residue (together with the cereal stubble) should prove sufficient to maintain both adult and young sheep in good condition – although caution should be taken to see that not all the medic seed is consumed. As he points out, under the existing grazing regime, once the grain component of stubble is eaten the remaining cereal straw is not sufficient to maintain the sheep and many lose weight – in particular the young sheep, which have difficulty in digesting the high fibre straw. The additional source of dry medic residue would prevent this loss of weight which leads to death in some cases. In many years the residues of medic and cereal crop should be sufficient to maintain the flock until 'well into September'. The suggestion is made that one could sell off the young sheep early because they had grown so rapidly so that the declining amount of fodder available at that time can be reserved for the breeding nucleus of the flock.

From the middle of September to the onset of the first winter rain, when the medic will regenerate, supplementary sources of feed should be kept on hand. It is suggested that a store of grain or hay (or a combination of both) should be kept. A little barley may be grown and the grain harvested for use during this period (Pattison, 1978, p. 23). There is an advantage in feeding this directly in the paddock itself as it encourages sheep to forage for whatever plants or residues remain on the ground.

Naturally November is the most difficult period for feed supplies. If the rains come they spoil what dry matter remains, and yet it can be several weeks before the new pasture is available. In addition, the cereal rotation must be begun on the other half of the farm, and this means the cultivation of the soil and the loss to the livestock of half of the remaining residues. The stickiness of the soil following the rains makes it impractical to have the sheep picking for what green seedlings may emerge, and hay and grain should be fed in the shed. Once the rain has fallen it is not long before the four to six leaf stage of the regenerated medic pasture is reached and at that time the sheep should be re-introduced to the field, although some hay or grain may be needed if pasture growth is slow due to cold conditions. Usually the medic pasture is able to support the livestock again by late January.

Nutritive value of medic

The trials carried out by the team in Algeria showed the medic to be high in protein and of good digestibility. Although the feed value varied throughout the growing season 'at all stages [medic] is superior to that of gramineous pasture plants or weeds' and Pattison provides details of research work by Adem and Vercoe and Pearce carried out in 1969 to substantiate his point (Pattison, 1978, p. 11).

The nutritive value of the dry medic residue compared to that of pasture of a gramineous nature should not be overlooked when comparing the benefits of each. Not only does medic share with other leguminous pastures a relatively high protein content in the dry leaf, but to this must be added the high protein content of the seed which can have at senescence a crude protein content of 15%. Naturally this deteriorates with time, but dry medic is 'always superior to the other dry feeds available' (Pattison, 1978, p. 12).

Provided the balance between sheep and medic is correctly achieved, a quantity of 1 tonne (DM)/ha of dry residue is likely to be available at plant senescence together with 200–300 kg of seed. The net result is that not only can the livestock be maintained in good condition on the medic residue, but that older sheep can actually be fattened. At least 20 kg/ha of medic seed will need to remain in the soil to supply the reserve from which the subsequent pasture will regenerate. Normally one would recommend that at least 50 kg/ha be retained to ensure that there is ample seed to cope with false breaks in the season or inappropriate cultivation and so on (Chatterton & Chatterton, 1989, p. 52).

Replacement of hay and straw by dry medic pods and straw

Residues of grain and straw are still the single major local source of the feed for Algerian livestock between July and October, but Pattison predicted that the availability of dry medic pods and straw would provide a significant increase in the supply of cheap, good quality grazing during this period.

The benefit of summer grazing for the farmer is that there is no need to make hay or store it or feed it out later – all time consuming and costly operations. Summer grazing of pods and straw pasture residue on the ground is almost unknown throughout the region although it is no different to the grazing of cereal stubbles.

Hay and silage (widely used in the higher rainfall zone in Algeria) seemed to Pattison to need relegation to the status of 'supplements to make good the differences between feed demand and supply'. He considered that while the land preparation for the conventional hay mixture of oats and vetch is less than for wheat, 'quite high costs of cultivation and sowing are incurred together with seed costs', and he pointed out that once medic is well established, medic hay can provide a cheaper and more nutritious replacement. Comparative figures for quality are medic hay 15% crude protein and oats and vetch hay 10% crude protein. The quantity also favours medic. Pattison quotes 4/5 tonnes/ha of medic in an average season (e.g. 1976) as against 2.48 tonnes/ha from oats/vetch (Pattison, 1978, p. 16).

Medic pasture, fodder crops and fodder conservation

During the life of the project a great deal of work was done to test the adequacy of medic pastures as a reliable replacement for fodder crops, hay, grain and volunteer pasture on the farms on which the teams established and exploited the pasture.

Fodder crops could provide a 'bulk of early feed which could be grazed before medic pastures are ready for stocking'. The qualification was that 'A fodder crop which does not produce grazeable feed early in the growing season would be of little use in increasing animal production regardless of how much feed it could produce later in the season' (Pattison, 1978, p. 13). The problem is that the farmer would have to sow early to obtain early feed and this would mean he would have direct competition for the time and resources he would be using to undertake the sowing program for his cereals.

Pattison suggested that irrigated lucerne, if water is available, and dry if it is not, could be useful, but cautioned that 'lucerne cannot be expected to make growth throughout the whole of the summer'. Lucerne will only grow well if 'carefully managed and grazed on a rotational basis'. It would provide green feed for about six weeks after other pastures had dried off. It was noted, however, that at the time the report was written most lucerne varieties in Algeria had been developed for cutting and that to proceed along the lines recommended it would first be necessary to find and develop grazing varieties.

The mechanics of comparative measurement of available feed

Pattison questioned the use of cut and measure as a means of assessing the productivity of medic versus forage crops. If a single cut is made during the peak growing period of forage crop and medic plant and the yield is weighed, the assessment can be that medic is less productive than for instance, vetch; but this discounts completely the economic value of medic that provides constant grazing of green pasture re-growth and later the dry pods and straw that remain on and in the soil to serve as dry fodder, which is available before and after the stubble from the cereal harvest.

Such a simplistic comparison also takes no account of the production of seed for the subsequent regeneration of the pasture or the continuing enrichment of the soil from the grazing that occurs.

H.C. Trumble, in his account of his own career as a grassland specialist, refers to English pasture specialist, R.J. Stapledon, who pointed out in the early 1920s that plants only become a pasture after they have been grazed and that the value of the pasture can only be measured realistically after trampling by animals and the return of their urine and faeces to the pasture. Cutting a trial plot is not a sufficient measure of the value of the pasture (Trumble, 1948, pp. 46–7).

Growth rates of livestock on medic

Although data on the productivity of sheep grazing medic pastures in Algeria were meagre at the time the livestock expert wrote his report, he was able to quote a demonstration trial at Hammam Bou Hadjar where liveweight gains in lambs of two to four months of age (still suckling) was 228 g/day when stocked on medic pastures at a rate of 10 ewes plus lambs/ha. This gain was comparable with those obtained with concentrate feed rations (Pattison, 1978, p. 12). The advantage for the farmer with the medic is, of course, that the cost and labour involved in its production and exploitation is considerably less than that involved in the production and exploitation of forage crops, and the cost is much less than that of purchased concentrates.

Nomadism

The large migrations of sheep and other livestock from the desert to the cereal zone that take place in Algeria (transhumance) in summer seemed not to unduly bother the project team.

They did not have significant problems with nomads illegally grazing flocks on project farms (certainly nothing as traumatic as those experienced by the demonstration farm team in Libya), but Pattison refers to the possibility of unauthorised incursions onto pasture land and he speculated about possible changes in management practices in the cereal zone leading to a diminution in summer grazing for rangeland-based sheep and whether this would cause hardship to owners of sheep in the rangeland.

There was a longstanding concern within the Algerian Government about the dissatisfaction of nomads due to the deteriorating condition of the rangeland. 'La Problème de la Steppe' provided the focus for countless conferences, workshops, seminars and learned – and not-so learned – papers from Algerians as well as foreigners. Pattison relates that

... the problem of transhumance is already recognised at Government level and alterations as proposed in the Wilaya of Saida (and possibly other Wilayas) are steps in the right direction. These changes which involve trying to settle the sheep owners on specific areas, providing all the facilities needed for living, controlling grazing etc., are all part of giving these people a better life and at the same time helping to restore the steppe. These programmes have been developed without the thoughts of the change in grazing pattern which would occur with the widespread introduction of medic pastures. However they would fit in well with what is required if a medic programme is to succeed (Pattison, 1978, p. 35).

A proposal for a second phase of the project

Pattison recommended that a second phase to the project should be funded and that it should concentrate on adequately demonstrating the whole farming system of medic in rotation with cereals. He was confident that an increase of 50% in livestock numbers above the mean could be obtained if medic pastures were introduced onto cereal farms and managed properly (Pattison, 1978, pp. 32–5).

He strongly recommended a great increase in extension information, and an intensive training course in the system for the technicians who were responsible for guiding farmers to adopt it. He believed that it was important to have practical demonstrations of the entire management system on 'at least 100 hectares', as well as the importation of 'suitable equipment' – i.e. scarifiers and combine seeders suitable for a system of shallow cultivation.

Pattison argued for a concentration on action rather than pure research. *A propos* of grazing trial demonstrations he writes 'Detailed stocking rate trials especially those featuring replications are not warranted . . . The functions of . . . demonstrations would be . . . to show farmers how to manage pastures. The effects of mismanagement should be demonstrated and pointed out'. He revealed a little of his own Australian conditioning by recommending that 'fencing would be essential' although he must by then have seen the control exerted over flocks by shepherds. He is brusque in his recommendation that sowings of medics should be

restricted to farms which:
(i)   will intelligently follow instructions on cropping and grazing techniques;
(ii) have sufficient sheep available to effectively graze the medic pasture;
(iii) have suitable cultivating equipment available (Pattison, 1978, p. 72).

Extension agents and training

We found no account of formal training programs undertaken by project staff for technicians or farmers although it was the original intention that the agronomists working in each region should train an Algerian counterpart and carry out such programs. The project did leave behind in Algeria a small group of Algerians who worked at various times with the project, some as technical assistants, and who were convinced that a medic/cereal rotation would fulfil the potential forecast for it in Algeria. Several went on to gain academic qualifications at Australian institutions and others published papers on the use of medic pasture in Algeria. A small number of diplomates from Algerian institutes went to South Australia to study for a year at Roseworthy Agricultural College when it established a Diploma in Dryland Farming, but few found career opportunities to use this expertise in Algeria.

A review of the project

In spite of the fact that the team went directly into existing Ministry-controlled and financed farms and sowed medic on 30 000 ha, only one farm manager (at Domaine Chouhada) successfully established and then continued a medic/cereal rotation, and even that was a kind of curiosity – only 250 ha out of the 3740 ha of the total farm.

The project provided the Algerian Ministry and technical institutions with strong evidence that the integration of medic pastures into existing farms could result in better soil fertility and increased production of livestock.

The project team were hampered by the fact that the architects of the project in FAO saw the medic pasture simply as another legume crop to be substituted for fallow to increase cereal yields. Because of this they prepared a machinery list that reflected the needs of deep ploughing and not shallow cultivation.

Was it ignorance of the requirements of the medic system that caused them to order the inappropriate machinery?

Were they influenced by Doolette's belief that the ordinary farm equipment on hand could be used adequately for shallow tillage if it was 'adapted'?

This barrier to a successful rotation allowed weeds to become a problem in the cereal phase and inhibited pasture regeneration. The advantages to the farmer of reduced cost, reduced time taken, and the effect of the level seed bed on the efficiency of other operations such as herbicide application and harvesting, were not demonstrated.

Lack of experience in grazing medic pasture and the lack of sufficient sheep to achieve good management meant that the pasture was never properly exploited and its full potential unrealised. The management regime sketched in the livestock expert's report could provide a guide to the Algerian dryland farmer but it remains unread in FAO archives in Rome.

The opinion of the Australian consultant agronomist that the local varieties of medic would flourish voluntarily if superphosphate was applied before the autumn rains could have been a useful means of expanding the pastures. There is no evidence that this was tried.

Those who planned the project directed that technicians working with the team should be taught the underlying theory and practical operations of the farming system but did not appreciate the need for skilled demonstrations of the management of the system – the different type of tillage and the managed grazing – that the farmers would have to adopt if they were to successfully operate a medic/cereal rotation.

The livestock expert wrote 'Sufficient knowledge is already available to make ley farming work if the farmers can be reached, instructed and convinced of the advantages of the system' (Pattison, 1978, p. 72), but where in Algeria was this 'sufficient knowledge' available? Certainly not at the level of the technical adviser and even less so at the level of the farmer.

The absence of shallow cultivation in the cereal phase prevented a useful demonstration of the interlocking nature of the cereal and medic phases and farmers did not see the direct relationship between the medic regenerating and providing them with feed for their sheep with little effort and outlay. The State and Cooperative farms on which parts of the system were demonstrated had no interest in any savings that may occur and were shielded from any benefit that greater profits may bring.

Perhaps the Algerian farmer will eventually be able to develop his own medic-based system, but how can he do that without falling foul of the interventionist technical experts who control most of his management for him? As we shall see below, things have changed and more farmers have independent farms, but they are still being propagandised and enticed to follow the old French-inspired system – which is the contrary of a medic based system.

Extension material

Extension material was obviously needed to counteract this. In 1978 the South Australian Government published a French translation of *Farming Systems of South Australia* (Webber, *et al.*, 1976) and a film called *Food from the Reluctant Earth*. These both described the system in the context of the South Australian farms and did not attempt to deal with the particular problems of, for example, deep ploughing. Until this material was published there was no comprehensive account of the South Australian medic system available anywhere in the world. This material was widely distributed in Algeria. Later an audio-visual kit in French and Arabic still on the Australian version of the system, but with a small section showing the danger of deep ploughing, was made available (Chatterton & Marsh, 1980). These still did not provide the demonstration necessary to the Algerian farmer who wanted to know how deep to plant medic seed, how often to graze it, or where to buy a scarifier, what he would save and what he would receive in return.

The aftermath

The Algerian project depended on technical research leading the way. The decision to go straight in and sow medic pasture on a number of large farms was taken with confidence but the lack of appropriate machinery, the absence of experienced practitioners of the system, and the short term of the project combined to leave behind very little in the way of radical or widespread change.

The project was not without positive results. It confirmed what Doolette had found in Tunisia – that deep ploughing was not compatible with the medic system and that the Australian cultivars could be used to good effect in the cereal zone. The Algerian project demonstrated that while medic was a good green manure for cereals, it had even more value as good quality feed for sheep. The project also showed that equipment designed for deep ploughing could not be successfully 'adapted' to the operation of shallow tillage.

A few useful observations were made, among them that local medics may well respond to superphosphate sufficiently well to make heavy initial sowings unnecessary. Another was the way in which the local ecotypes soon re-established and then dominated introduced cultivars once their seed ceased to be ploughed in deeply during the cereal phase.

Although there was no second phase of the project, the Algerian Ministry did not lose interest in the ideas introduced by the team but they were left to rely on a basket of technical reports for their further expertise. Problems had been identified, potential sketched, but the system was not in operation on the farms or on the research stations.

Instead of providing scarifiers to the farmers and bringing in a team of expert medic farmers to teach Algerian farmers how to manage a medic system, the Algerian Government decided that a reorganisation of the administration and its associated institutions was necessary as a first step in preparing the way for farming change.

(a) Institutional reorganisation

The attempt to clear away the excessive bureaucracy and the agrocratic didactism that pervaded the agricultural sector took place under Minister of Agriculture Selim Saadi who prepared plans for a mammoth reorganisation between 1979 and 1983. He was interested in the medic/cereal system and he encouraged his officials to take an interest in it.

Saadi set up a High Commission for the Steppe (rangeland) to try to discover dynamic and innovative ways of carrying out revegetation of the rangeland to save it from the increasing desertification that threatened. The High Commission for the Steppe began to show promise of achieving its objective with a vigorous examination of rangeland reclamation options, including a study of the potential of medic to provide pasture and reduce erosion (Tebessa Seminar, 1986).

In the early 1980s a rangeland rehabilitation project at Ksar-Chellala, operated by an Australian agency, briefly considered using medics to increase livestock feed over a large area, but the project failed to progress beyond the mapping and planning stage (K.C., 1983, Final report).

In a separate development, the Ministry of Forests (particularly under the Directorship of the energetic and innovative Abdellaoui who had visited South Australia as Director of Forests and was later appointed Minister) carried out reforestation programs in hill country and included in some of the higher rainfall areas a medic component to provide under-forest grazing (Chatterton & Chatterton, 1979–90).

In spite of all the good intentions, none of these reorganisations or initiatives brought Algerian farmers nearer an improvement in their farming systems.

Domaine Chouhada

In Libya in 1983, there were over 600 farms in the El Marj district where farmers were using a medic/cereal rotation. In Algeria in 1983, the only farm (Domaine Chouhada) to continue the medic/cereal rotation introduced in the early 1970s was struggling to keep it.

We were able to inspect Domaine Chouhada at Khemis Miliana, in 1979, 1980 and 1983. The Domaine had 3740 ha of arable land and 120 workers, and it averaged 375–400 mm rainfall annually. The farmer-manager (Chairman of the cooperative that owned the Domaine) had (initially with some caution) set aside 250 ha for a medic/cereal rotation as his part in the FAO medic/cereal project and he personally carried out much of the work involved. He said that the number of sheep on the farm had been increased from 200 to 2200 (all maintained on medic) and in 1980 three shepherds were employed to guard them. By 1980 he had pasture to spare but he could not expand his sheep flock, firstly because of difficulties in buying in extra sheep and secondly he was reluctant to incur the capital cost required to erect further bergeries (sheep sheds). These are costly concrete structures in which the sheep are herded at night. The farm manager said that these were essential as he did not believe that the sheep would survive if left out at night due to the cold in winter. He found it hard to believe that sheep are left out all night in Australia in conditions as cold and colder, and that it is considered that sheep kept in sheds are liable to pneumonia and other diseases.

The medic pasture on the farm (rapidly dominated by local ecotypes re-establishing in the favourable conditions) regenerated without difficulty because he used a scarifier for shallow cultivation during the cereal phase and managed his grazing well. The increased soil fertility that had built up over the years following the introduction of medic led to an increase of weeds in the cereal crop and because of this yields were for a time a little less than those on land where bare fallow was still being used. The manager had persisted with the rotation in spite of this because of the value of the pasture to his livestock. In 1980 he was able to get supplies of a pre-emergent herbicide and used it for the cereal crop, which we observed to be weed-free. It out-yielded the post-fallow crop.

Domaine Chouhada was on the transhumance route and each summer tribal flocks came to graze the cereal stubble. The manager reported no difficulties in keeping these flocks off the dry medic, and they usually left before the medic pasture regenerated following the autumn rains.

While the farm manager did not refer to the lower costs associated with the medic/cereal rotation versus the traditional deep ploughing and bare fallow as a reason for his persistence with it, he said that it was due to the profitability of the livestock on the medic pasture that enabled the Domaine to make so much profit that it was able, at last, to have 'its own chequebook' and not have to go to the Ministry for approval and disbursement of running costs.

In 1983 the Domaine had a bad infestation in the cereal crops of the common 'vers blanc' (white worm). The MARA agronomist attached to the Domaine blamed the introduction of medic for it and was sceptical when Australian experts (familiar with it on Australian farms) said that it came with the cereals and not with the medic. The advice for treatment given by MARA (Anon., 1983) was unnecessarily costly and not undertaken and the effect of the 'vers blanc' on the medic was devastating. An entomologist was sent from South Australia to provide advice (Birks, 1983) but the MARA agronomist had, in the meantime, insisted that all the medic should be ploughed up and the entire farm revert to the old method of deep ploughing and fallow. We have no later information on the farm.

Conclusion

Algeria's average yield for cereals remained at 730 kg/ha (FAO Production Yearbook, 1991) and the ratio between prices of sheep meat and barley widened from 30:1 in 1970 to 66:1 in 1987. Overgrazing of the rangeland reduced the carrying capacity of the range by half and it is now only capable of providing about 10% of food for the animals kept on it (Treacher, 1990). In spite of subsidies for livestock production, by 1988 the price per kilogram of sheepmeat to the consumer was 70 AD (US $20). The average Algerian family considered itself lucky to eat mutton once a week.

From the end of the project until 1983, opinion about a medic/cereal rotation split into two camps in Algeria. One group said that the 'Australian cultivars' were not suitable to Algeria and that until Algeria found its own cultivars, the medic on Algerian farms would fail. Another group remained convinced that the medic (even if grown from Australian cultivars) would help to solve the problems of declining soil fertility and ever more costly livestock production. The question was, how to go about it?

The Algerian Government followed what became the conventional path of technical and scientific progress towards a medic system. They funded a nation-wide research program to identify local cultivars (with the assistance of ICARDA and ACSAD in Syria) but did nothing directly on farms to try to apply the findings of the FAO project. They did import two hundred combine seeders from Australia and distributed them to research centres and state farms. They were unable to also acquire expert farmers to demonstrate their operation and maintenance and they remained unused.

The FAO/Algerian medic project did find useful things about the use of medic on farms of varying sizes and rainfall, but this information remained locked in reports. The critical transfer from the project sites to the neighbouring farms did not take place. As the Libyan projects demonstrated, farmers when provided with the tools to undertake the mechanics of the change and provided with proof of more production and better profits do adopt the new techniques and management of a new system on their own farms. But the Algerian farmers did not get such assistance.

Without this, technicians and farmers alike were left floundering in the dark after the project team left.

7

A medic project in Jordan

Introduction

By 1979 the Libyan projects were providing good models of the cereal phase, the grazing phase, and rangeland re-vegetation using medics. The provision of appropriate machinery and expert farming advice to farmers was achieving rapid adoption of these models.

The Algerian project had finished and had left on record further proof that shallow cultivation and regeneration of medic pastures depended on the availability of scarifiers and that medic pastures provided cheaper and more sustainable grazing and nutrition than was presently available in the region. The Algerian project also showed that without direct support to the farmers adoption of project results could not take place.

Neither project had encountered difficulties using the imported medic seed from Australia and both had found large numbers of local medics in many locations that could deal with possible problems of climatic or economic origin. Neither project had established a specific seed production program but had made a good start on the identification of local ecotypes and this together with shallow cultivation and precision seeding made up the package necessary for a successful seed industry.

The achievements of these projects created much interest in the region and the Jordanian Ministry, already negotiating with the Australian Government aid agency (AIDAB) for an agricultural project, was keen to try a medic system on their farms.

The Jordan project began in 1979 and continued until 1991 when the war in Iraq caused the evacuation of the team.

Jordan's dryland farming zone

The staples of the Jordanian diet – pitta bread, mensab (boiled lamb) and labn (sheep's milk yoghurt) – come from wheat and sheep which are for the

166

most part produced on dryland farms. In common with other countries in the Near East, Jordan is increasingly reliant on imported sheep meat and grain, the latter not only for human consumption but also to feed livestock because of the use of bare fallow and the rapidly disappearing pasture on the rangeland. Between 1965 and 1971 imports of grain to the region increased from 4.4 million tons to 10 million tons (El Ghonemy, 1980, p. 4).

Ninetyone percent of Jordan is mostly rangeland and barren desert with an average rainfall of less than 200 mm, only 6% has an average of between 200–350 mm, and only 2.5% has between 350–800 mm. In the eyes of most experts from the Northern Hemisphere only the latter is considered reliable wheat growing country (El-Sakit, 1979, p. 2).

Irrigation

Jordan's horticulturalists in the irrigated Jordan Valley have for many years been the main recipients of credit and government encouragement to produce irrigated fruit and vegetables not only for domestic consumption but for what were rapidly growing export markets in the Gulf. The war with Israel in 1967 and the loss of a large part of the West Bank left the Jordanians with a decreasing source of irrigation water, and in 1980 a new dam was being planned for the River Yarmouk to provide a more secure water source. In 1987 the existing neighbouring markets (Kuwait, Iraq and the Gulf States) were oversupplied with fruit and vegetables and growers who had expanded production were reeling from heavy losses, their plastic houses empty and they themselves debt ridden (Chatterton & Chatterton, 1986, 1987 and 1988). The finance for the proposed dam was withdrawn in the early 1990s.

Jordan's dryland farming system

Many farms are small, and the smallest are 'strips' that are only a few metres wide and a couple of hundred of metres long. Fiftyfive percent of the farms are of less than 50 dunums (5 ha) but around another 30% are between 50 and 150 dunums (15 ha). Fifteen percent are large and in private hands. Many small landholders do not live on their farms and get only between 10 and 20% of their income from the land. Twentythree percent of the farms are sharecropped – the owner sharing the income in return for a neighbouring landowner working the land (Peckover, 1984, pp. 1, 2 and 3). An historic profile of the larger farms where sheep and cereals are produced can be gained from Raouf Sa'd Abujaber's account of Transjordanian agricultural activities (Sa'd Abujaber, 1989).

Jordan's dryland farmers usually count on a good cereal harvest in only one year out of every five because of unreliable rainfall and the decreasing fertility of the soil (Heysen, 1979, pp. 7–8). Nothing can be done about the rainfall but an attempt has been made to introduce nitrogen fertiliser as a means of providing soil fertility. It has not been widely successful because of variable responses obtained and because farmers are not prepared to expose themselves to the economic risks associated with the use of nitrogen fertiliser in conditions of unreliable rainfall (El-Hurani, 1980, p. 86).

Most farmers in the cereal zone, as well as those in the larger but more fragile marginal zone, sow wheat after wheat or after bare fallow. In the higher rainfall districts (around Madaba, Salt and Irbid) instead of fallow, they grow summer crops (melons usually) and tobacco (although the government began to discourage it in 1987) and rarely use fertiliser.

Many farmers keep a flock of sheep to provide meat for the family and cash from the sale of lambs. Sheep graze the cereal stubble and roadside weeds, but for the most part they are fed tibben (wheat straw chopped up, but in which there is a fair percentage of grain due to inefficient grain harvesting), baled straw and some alfalfa hay. Pasture is not grown.

The use of post-emergent herbicides for weeds in cereal crops has been encouraged by the Ministry of Agriculture and this, by diminishing the seed bank of adventitious plants, has helped to eradicate what little pasture existed on the rest fallow in the past. In order to speed the adoption of herbicides the government made finance available at advantageous terms to farmers so that they could buy spray plants (El-Nabulsi, 1980, pp. 97–9).

Mechanisation of farm operations has only penetrated 45% of the larger farms. In 1981 there were 5740 tractors in Jordan (Peckover, 1984, p. 7). Most tillage and harvesting with machines is carried out by contractors – the largest being the Jordan Cooperative Organisation. Eightytwo percent of the farmers who use mechanised cultivation do so through the JCO or other small private contractors (Peckover, 1984, p. 3). The 8.3% who use mechanised seeding use the same contractors (Peckover, 1984, p. 18). Since 1950 the contractors have used a three disc plough, three furrow mouldboard plough, six to eight disc harrows or a locally made rigid tyne cultivator two metres wide to cultivate for cereals.

It has been said that 'The three disc plough and the manner in which it is used has been responsible for widespread erosion and land degradation in Jordan' (Peckover, 1984, p. 3).

Farmers without access to mechanised cultivation, plough their fields using a steel tipped wooden plough and a mule, broadcast their seed and fertiliser (if used) by hand, and most harvest by hand.

That the existing system of cereal production is a dismal failure can be seen in the recorded wheat yield of Jordanian farmers which is one of the lowest in the world with a sixteen year average of 560 kg/ha. Attempts before 1980 to improve production by programs aimed at introducing and diffusing improved technology have been of 'limited success' (El-Hurani, 1980, p. 92).

The adoption of a ploughed bare fallow has increased the tendency to erosion. On the degraded farmlands near Zarqa and Mafraq in the north and on the dry side of Karak, Tafila and Shaubak in the south one sees the farming of desperation – tractors ploughing up soil that has no fertility and precious little structure in the forlorn hope of reaping a poor crop or at least getting a little grazing for sheep to augment the purchased tibben and straw.

Livestock and the rangeland

There is interdependence between the cereal zone and the rangeland because some village families have rights to rangeland pastures as well as access to village commons, and rangeland flocks graze the cereal stubble after the harvest. Most livestock owners have traditional rights to areas of rangeland but do not own the land.

The rangeland where the larger flocks are based is bare, devoid of trees and grass. Attempts made to revegetate it by either removing livestock completely to allow natural revegetation or by planting *atriplex vesicaria* and other species and keeping stock off it have not succeeded in producing more feed because in both cases the vital component is the removal of the livestock from the site.

Nothing has been done to solve the problem of income for the stockowners during the long periods of exclusion, nor, it must be said, have these schemes increased available pasture when opened up. The atriplex has proved just as incapable of sustaining intensive grazing in Jordan as it has in Australia (Squires, 1981), and there has been a growing reluctance even to open up rangeland revegetation perimeters to livestock. One station in the vicinity of Ramtha has now been fenced off since 1932. It has a fine regrowth of natural vegetation on it but no stock are allowed to enter for fear of overgrazing (Chatterton & Chatterton, 1979, 1986 and 1988).

The decision about what vegetation should be used to restore the rangeland has been made by rangeland experts who favour using fodder shrubs, mostly *atriplex* spp., but the method used to establish them is costly. Atriplex plants are raised in nurseries and then planted out by hand into either ripped rangeland or in holes dug for each individual plant.

Water is carted to the site and each plant watered individually until they either establish or die.

Cactus has been used but this also needs scarce water to establish and has not proved its worth in rangeland areas (FAO, 1985, p. 27).

In effect, these schemes have reduced the amount of grazing available to rangeland flocks, yet one sees them right through the Middle East and the Maghreb.

The continuing degradation of the rangeland is forcing livestock owners to depend almost totally on purchased feed for their stocks. Fodder produced in Jordan is for the most part tibben and some alfalfa hay, supplies of which are not enough to be significant. Some uncultivated fallow is available for grazing and this supplies about 14% of the feed requirements of livestock. It was estimated in the 1980s that 75% of all fodder was imported. Imports of common feedstuffs to Jordan increased from 45 017 tonnes in 1971–3 to 235 703 tonnes in 1981 (Qureshi, 1984, Table 5).

Conventional solutions to the problems

The problems encountered by Jordan's cereal farmers and livestock owners have not lacked attention from the international agencies. For instance, the Ford Foundation's Middle East Regional Office published a series of essays by technical experts familiar with the region about the problems of low production and environmental damage on dryland farms.

Most of the authors referred to the need to replace bare fallow with legumes, warned against the effects of deep ploughing and inefficient tillage, provided examples of the failure to persuade farmers to change from a weedy fallow on which they pastured their sheep to a cultivated bare fallow that was (inaccurately so) credited with conserving moisture for a subsequent cereal crop, and pointed out that it was necessary to overcome the increasing denudation of the rangeland (Arer, 1980, pp. 18–19).

The conclusion of these experts was that solutions could be found if more research programs were undertaken to collect more data (Gotsch, 1980, pp. 1–5 and 134–8).

The essays left the reader with an impression that dryland farming in the Near East was a black hole into which experts threw bits of farm technology from here and there and hoped that something useful would emerge.

Dr Haitham El-Hurani, searching a little wider for possible reasons for the failure of many past development projects in the region, suggested that there may be 'some sort of gap between what is known by professional agricultural leaders and researchers and what is perceived by farmers who are the ultimate users of these technological changes' (El-Hurani, 1980, pp. 81–92).

Medic as a solution

Jordan has had a rich supply of indigenous medic ecotypes. The depradations of continued deep ploughing and overgrazing have caused most of it to disappear.

If medic pasture can be reintroduced onto cereal farms to replace bare fallow in the rotation as it has been in Libya, it would increase soil fertility and improve both the profit and productivity of the farmer. If farmers either increased their own flocks or rented pasture to rangeland flocks, this would take pressure off the overgrazed rangeland. If, at the same time, medic pasture could be re-established on large areas of rangeland, as it was by the Western Australian team in Libya, and a grazing regime agreed with livestock owners, the increase in available grazing there would lessen the dependence on imported feedstuffs and help reduce the erosion of the soil (Chatterton & Chatterton, 1984, pp. 117–29; Carter, 1975).

The Jordan/Australian dryland farming project

The original intention of this project was to emulate the Seedco project in Libya. The project would buy seeds and machinery and employ farmer experts to demonstrate directly to Jordanian farmers the skills needed to operate a medic/cereal rotation. It was to be, in effect, a strong extension effort backed by the right tools to do the job. The demonstrations of the new system would take place on existing small farms and the target would be the poorest farmers. The project sites are shown in Fig. 7.1.

The partner

The Jordan Cooperative Organisation (JCO) was appointed a partner with the Australian project team. There were several reasons for this decision.

- The JCO carried out most of the contract services for cultivation and seed bed preparation and would be the means of introducing shallow cultivation using scarifiers and combine seeders to Jordan.
- The JCO contract tillage program had established working agreements with cooperative groups of small landholders (particularly those belonging

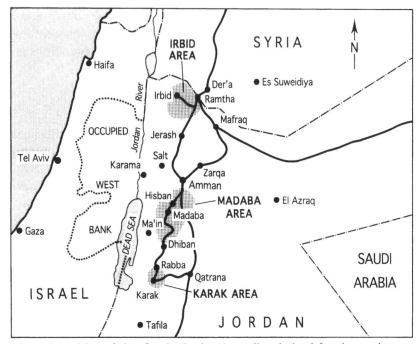

Figure 7.1.   Map of sites for the Jordan/Australian dryland farming project.
(Source: SAGRIC International.)

to individual families) so that their adjoining strips of land could be
cultivated, sprayed and seeded as one. It was hoped that this form of
cooperative management could be extended to cover demonstrations of
a medic/cereal rotation on these very small holdings.
- The JCO employed a team of field officers who were technically qualified
  to provide a farm extension and advisory service and this was considered
  a potentially useful medium through which to disseminate information
  about the establishment and use of medic. Several JCO technicians were
  selected to become counterparts to the project.

The project team

The original intention to have a strong component of expert medic farmers
as principals within the project did not eventuate. The South Australian
Department of Agriculture's new consultancy unit, SAGRIC International,
was awarded management of the project by AIDAB and they chose the team.
   The first team consisted of an agronomist, an agricultural economist,

and a farm machinery expert who was fortuitously an expert medic farmer. He was employed to carry out the tillage operations on farms involved in the project, as well as assessing the feasibility of a proposed machinery centre at Madaba to be funded by the German aid agency GTZ. He worked from within the JCO and, in the initial stages, trained and monitored their operators in the use of scarifiers and combine seeders to prepare seed beds and sow crops.

During the first phase several part-time consultants from Australia were employed as technical experts by the project. Later teams consisted of technical experts (including livestock experts) and two farmers who were employed as tractor drivers solely for the project's programs.

The project in operation

The target farmers

The intention to concentrate on the poorest farmers did not survive the original declaration. In their place, farmers of moderate means were found who were interested in joining the project and trying the medic on their farms.

Because of the late arrival of the team, no farmers were involved in the first year's program, but in the second year the team worked with some farmers to sow medic on part of their farmland. They were to be shown how to graze it and, later, how to cultivate the land in preparation for a cereal crop without endangering the regenerative capacity of the medic. The project provided the seed and fertiliser and put a perimeter fence around the medic pasture to keep out unauthorised sheep, as well as subsidising contract charges for tillage and sowing operations carried out by JCO. The farmer, in his turn, had to agree to be directed in the management of the fields he had contracted to the project and to provide labour as required (JDFP, 1984, 1 (15) Appendix 5, p. 2). As we shall see, the intention to sow pure medic pasture lasted little longer than the intention to concentrate on the poorest farmers.

Over time the team changed their perception of the role of the farmers *vis-à-vis* the project. In Annexe B of the first *Quarterly report*, the objective states that: 'From the early stages of the project, Jordanian farmers will be involved in all aspects of the operation, including decision making' (JDFP, 1981, 1 (1) Annexe B, p. 11).

Annexe C of the same report went into some detail concerning the project's responsibility to the farmers. The team would be aiming at the creation of an 'attitudinal change' on the part of the farmer from one of 'risk minimisation' to one of 'profit maximisation'. The project would 'rely largely on an intensive extension effort in order to gain farmer acceptance of new technology' and in particular to increase farmer skills in 'rational farm management decision making'. If rationality could be established as the criteria for farming decisions then it was hoped that the simplicity and profitability of the medic/cereal rotation would become apparent. The project team were to act 'purely as guides and advisors – all work should be carried out by farmers and as many decisions as possible should be made by them' (JDFP, 1981, 1, Annexe B).

Extension programs for farmers were slow to begin. The first field day was not held until 18 March 1981 and the majority of the audience were interested parties from the University, the Ministry, the JCO and 'several local farmers' (JDFP, 1981, 1 (1), p. 6).

A survey of farmers was undertaken early in the project in order to record profit margins and losses from haymaking and sold grazing resulting from the medic/forage program established on their farms. The survey was only attempted during one year and the team leader pointed out the dangers in drawing too many conclusions from it (JDFP, 1984, 1 (15), Appendix 4).

During 1982 a consultant, Dr P. Cocks, emphasised that the objective of the project was to establish a strong extension component to present to farmers. If this was to be achieved the team would have to first demonstrate to farmers that the medic system does, in fact, work. He believed that this was not being done (Cocks, 1982, p. 6).

An audio-visual representation of the project's activities and the operations it wished the farmers to undertake was made with the cooperation of the privately funded German aid organisation, the Neumann Foundation. The eighty slides covered shallow cultivation, haymaking, the use of contour banks to assist in the prevention of soil erosion, sowing and harvesting. The team leader wrote in 1984 that 'There have not been any showings to farmer groups as yet as it is important to gain the confidence of the research and extension people before going to the farmers' (JDFP, 1 (15), p. 4).

In September 1985 the team decided to divert their energies from advising farmers to the 'more formal activities' of training the Ministry of Agriculture and JCO staff. These more formal activities were to consist of field days, seminars and the recruitment of 'one or two Masters candidates to do their research work . . . and to receive post-graduate training' (JDFP, 1985, 2 (3), p. 6).

At the same time, the university made available a 20 ha site to the project and from that time work on medic pastures became concentrated there.

Technicians as counterparts

The decision to train technicians and select postgraduate candidates shifted the focus of the project from the farmers to the institutions.

At the start of the project, counterparts – technicians shown how to operate and manage the new system – were thought by the project planners to be the best way of providing a bridge between the foreign expert and the farming community and of leaving behind a source of support once the foreign expert had gone. 'Counterpart staff from the Ministry of Agriculture and the JCO will ... be trained to develop a full understanding of a pasture/cereal rotation and to obtain management skills' (JDFP, 1981, 1 (1), Annexe B, p. 11).

Unfortunately the JCO field officers were almost totally preoccupied with the selling and administration of credit, seeds, fertiliser and machinery services, and the team complained of their unavailability for training sessions.

Several technicians (from the JCO) did benefit from their early involvement with the project, but they were not then employed as extension advisers to farmers growing medic pastures.

It proved easier to find candidates to undertake postgraduate studies and several completed these. One of the successful candidates (having already gained his M.Sc.) was given a special dispensation to undertake a Ph.D. qualification that enabled him to spend half his time in Adelaide growing trial plots and attending classes at the Waite Institute and the other half in Amman growing and assessing trial plots on University land.

There was no attempt by the team to employ expert farmers with experience of extension work in Libya to fill the gap.

Trials of medic cultivars

The first sowings of imported cultivars of medic on local farms grew and produced well (JDFP, 1981, 1 (1)).

Trials for research purposes were not so trouble free. Much seed was sown late and although in the first season seed sown prior to 25 December showed 'excellent' growth, sowings in January and February were 'struggling'. In spite of this it was hoped that 'all sowings should achieve seed set and go into rotation next season'. This optimism was not reflected in results.

This first season was later described as 'a disaster as the only nursery grown that year was on farmer's land and was ploughed in during April'. Nonetheless it was concluded from the few plants that remained that *M. rotata* (an indigenous ecotype), *M. blanchiana* and *M. scutellata* were 'the most productive looking lines' (JDFP, 1984, 1 (15), p. 6).

In the second season there was a bad drought and only trials of forage crops are mentioned in the report.

The third season had exceptionally good rains and there was an excellent seed set and regeneration from *M. rotata* (the local ecotype) and three other lines.

In the fourth season (a very bad drought) a trial of inoculated medics was undertaken. All seed was inoculated with the rhizobium CC 169. The Ramtha trials were 'accidentally lost' just as they were recovering from the drought. Someone ploughed them in. From those trials that survived it was concluded that *M. rotata* had proven the best.

It was the agronomist's opinion that *M. rotata* should be recommended to farmers as a pasture in spite of the fact that it grew in an upright manner and produced its seeds high on the plant, 'thus making it susceptible to overgrazing'. An additional difficulty for farmers wanting to take this advice (not flagged in the report) was that the seed from this local variety was not available in sufficient quantities to meet the needs of farmers.

The agronomist noted that large seeded medics imported from Australia such as *M. scutellata*, cultivar Robinson, did relatively well but that the smaller seeded medics such as *M. truncatula* and *M. polymorpha* were difficult to establish in Jordan because of a tendency to sow them too deep and to lose them after the cereal crop. It was suggested that research be undertaken to identify a cultivar that could survive deep ploughing.

Given the 'accidents' attending much of this early trial work any conclusions offered about it must be considered speculative only.

In 1982 the visiting consultant (Dr Cocks) found it difficult to assess whether there were likely to be establishment problems in normal years because the previous season had been climatically difficult, but he was inclined to believe that inoculation was necessary for Australian cultivars used in Jordan – and possibly for any local lines that may be identified during trials. He recommended that all seed used in demonstration sowings should be inoculated with an appropriate rhizobium. The question of the need for inoculation loomed large in future research programs carried out by the team.

He also suggested that the project's practice of sowing the medic on farms at a rate of 4–5 kg/ha instead of the recommended rate of 10 kg/ha was creating a risk of weak pasture. He advised the team to use even higher rates of sowing, say 30 kg/ha, and that mixes of cultivars be sown instead of single lines (Cocks, 1982, pp. 5–9).

The poor growth of Australian cultivars on the research stations was often the result of poor selection of cultivars for particular sites, adverse seasonal conditions such as drought, poor seed bed preparation, and poor organisation of deliveries that meant that seed was often planted very late in the season.

At the Ramtha station, spontaneous medic was observed growing strongly on paths while trial sowings in plots either had not germinated or were just surviving. Questioning of the technicians responsible inevitably revealed some man-made reason for the failure on the trial plots, often the late arrival of the seed or an accidental ploughing. The technicians were surprised that the spontaneously regenerating medic growing between the failed plots could be taken seriously as an indication of its suitability to the climate in Jordan.

Later, when the 20 ha university site was used by the project for cereal production and grazing trials, the medics sown were Australian cultivars (among them cultivar Robinson) and they produced abundant feed and pods, and regenerated without difficulty.

Medic on the farms

In his summing up of the first five-year phase, the agronomist wrote that medic regeneration on farms after sufficient seed set and where no overgrazing had occurred was 'encouraging' and that 'initially medics can be grown productively in the marginal areas of the cereal zone'. He recommended that medics should be tried on 'stony hillsides and rocky outcrops' where cereal growing does not take place and, if successful 'their use will spread' (Bull, 1984, p. 2).

In total during the first five-year phase, 16 866 dunums (1686.6 ha) of farmland were cultivated under farmer contracts with the project (Peckover, 1984, p. 64). On average about thirty farmers (not always the same ones) were involved each year, although after 1984 farmers in the Madaba region (rainfall in excess of 300 mm) became less interested and did not fill the quota allowed them (JDFP, 1985, 2 (4); Harvey, 1984, 1 (15)).

Little attempt was made to sow pure medic on farms. Apart from one recorded sowing of about 28 ha in 1982, less than 5 ha was sown each year. In 1985 (at about the time the university site of 20 ha became available and after a Jordanian on the advisory committee for the project asked why pure medic was not being sown) the amount rose to 40 ha but declined thereafter (Reeves, 1993, p. 75).

Grazing of the medic

There may have been a little grazing of medic pasture in the first year on farms but it seems that most of it was sown in conjunction with other forages. Certainly hay containing medic was made (Harvey, 1984, p. 3). The agronomist writes of 'mixes' of forage being planted on most sites and of giving advice to World Food Program personnel on the planting of 2 ha of vicia (vetch) and barley for fodder (Bull, 1984, p. 2).

The final report for 1984 noted that one of the greatest failures of the project to that time had been the lack of 'A demonstration, WHICH HAS NOT BEEN ATTEMPTED AS YET, of sheep grazing in the field from the winter on' (Bull, 1984, p. 7).

In 1985 the farmer and livestock owner, A.R. El-Nabulsi, representing the JCO at a coordination committee meeting, made a special request to the project to plant 'medics as pure stands only without mixing it with other crops to find out its productivity and success in dry land areas with rainfall less than 300 mm. and on shallow soil . . . ' (JDFP, 1985, 2 (1 & 2), p. 17).

In 1986 a trial of lambs being fattened on a medic mix forage was held on Mr El-Nabulsi's farm, but his lateness in providing data made it impossible to insert an assessment in the project report.

The results of another grazing trial were affected by the fact that there were not enough sheep to form the acceptable social grouping defined by ICARDA as necessary, the sheep had not been pre-selected or pre-treated for disease and parasites and during the trial lost weight due to one or other of these factors and, because the whole program was late beginning, half the sheep had to be put on dry grazing which affected their comparative weight gain (JDFP, 1986, 2 (6)).

Between 1986 and 1989 a four-year trial was undertaken to compare liveweight gains of Awassi lambs fed beekia (vetch) in one group and medic (Snail) in another. The lambs fed on beekia gained more weight in the first two years, but when the Snail was replaced with what were called 'locally adapted cultivars' (*M. rigidula* and *M. rotata*) the lambs fed on medic gained more than those on beekia. This was taken to indicate the superiority of local versus Australian medics (ICARDA, 1989, p. 69).

Local accounts of the trial were that when the vetch was finished the lambs were removed from both fields. A farmer, when assessing the relative values of medic and beekia, would have taken the remaining dry medic feed as well as the seed bank for the germination of the next pasture into account. The costs and time associated with the need to re-sow the beekia and not the medic was not taken into account either.

In spite of all the opinions being put forward about the suitability or unsuitability of the medics to farms in Jordan, the only determined attempts to assess the grazing capacity of medic pasture took place on the site provided by the university.

Fences for grazing control

In Libya the team operating the demonstration farm feared invasions of unauthorised grazing. The team in Jordan insisted on fencing being erected before grazing on farms was attempted.

This proved difficult on the very small farms due to an 'apparent hesitancy [on the part of] farmers to group together cooperatively for agricultural operations'. The team, determined not to proceed with grazing programs unless the land was fenced, were faced with having to fence minute parcels of land if the hoped-for cooperation did not eventuate. In anticipation of this they ordered more fencing from Australia (JDFP, 1981, 1 (1)).

Hopes for the cooperative agreements were dashed, and eventually it was decided not to persist with trying to induce very small strip owners to come into the program. Instead, farmers with holdings of 10 ha and more were drawn into the project. The fear of unauthorised grazing persisted.

Some innovative methods were attempted in the hope of getting around the belief that all land was available for common grazing unless an identifiable crop was growing on it. In several cases, a harrow was taken over the ground during the summer to bury the dry pods and hide them from covetous shepherds and their sheep; in another, several rows of barley were planted around the perimeter of the new medic pasture to indicate the taboo of a growing crop. Eventually it was decided that the project would have to place 'more emphasis' on 'farmers who owned their own livestock or whose immediate family owned livestock which could be used to graze the forage grown at an agreed price' (JDFP, 1985, 2 (4), p. 3).

The fences that were erected were used by most farmers to enclose new plantings of olives and other fruit trees after the initial contract with the project had expired.

Hay or pasture

Almost from the beginning the team preferred sowing the medic as part of a mix from which to make better hay rather than demonstrating how to graze the medic pasture.

The expert medic farmer employed as the machinery expert had cast his eye over the farms and it was his opinion that farmers would profit more from grazing medic pastures than from trying to make hay from forage crops.

He reasoned that in the 200 and 300 mm rainfall zone a farmer could count on making good hay only in one year out of ten, but good grazing of medic pasture as well as a little hay making was likely to be possible 40% of the time, and a medic pasture could most probably be relied on to provide adequate grazing for livestock about 50% of the time (JDFP, 1984, 1 (15), p. 52). This cheap grazing, together with the storing of hay from the good years and the use of tibben and straw as supplement only, would enable the farmer to keep livestock at much less cost than under the existing system and would probably enable him to increase production due to the improved nutrition.

However, the decision was made by the project to concentrate on demonstrating to the farmers a forage mix and hay.

(a) Haymaking

Having sown mixes of medic, vetch and oats and found that these produced good forage, the team was initially very enthusiastic about the possibility of hay providing a good income for farmers who came into the project, besides avoiding all the knotty problems of managed grazing on small holdings and the fears of illegal poaching.

The first year that medic/vetch (beekia) hay was made, they were surprised to find that there was no market for it. It seemed that 'this is the first time that forage hay has ever been made in Jordan and no one will pay a decent price for it' (JDFP, 1984, 1 (15) App, p. 3). Lucerne (or alfalfa) hay was bringing 80 JD per tonne, but farmers who made 'forage' hay were only able to get 65 JD – not much more than their cost of production. This lack of interest was blamed on it being a 'good year' and therefore not much demand for forage hay.

The team believed the hay-buying public would have to be 'educated' about the value of forage hay and suggested an intense extension program be set up to achieve this. The machinery expert believed the government should subsidise the price of the hay to livestock owners and set up a special market for it (JDFP, 1984, 1 (15), pp. 54–5).

The continued under-utilisation by farmers of the haymaking equipment brought in for the project irritated the team members. During the first phase of the project the utilisation of the haymaking equipment was only 10–15% of its capacity (JDFP, 1984, 1 (15), p. 46).

Even more irritating was the presence of stones in the fields of those farmers who did join the program and the tendency of these farmers not to want to remove them. They caused severe damage to the haymaking equipment. Mechanised stone-pickers and rollers were imported to try to overcome the problem. Subsidies of 50% of the cost of hire were offered to farmers who wished to contract to clear their fields of stones, but few took advantage of them (JDFP, 1984, 1 (15), p. 27). The machinery expert who had to repair the machines thought that the project should refuse to sign a contract with prospective farmers unless it contained an agreement that the farmer would clear the stones from his field before the team came onto it. In the second phase of the project this requirement was incorporated into the contract but rarely honoured (JDFP, 1984, 1 (15), p. 57).

After several years of trying the team leader gave up in disgust any attempt to improve haymaking and to get the stones out of the field. 'What was apparent this year' he wrote, 'was that the art of hay making has not been understood by any of those involved in doing it. In only one case was decent hay made and that was more from good luck than good management' (JDFP, 1985, 2 (3), p. 15).

The whole program was a mess. The machines ordered by the team from Australia did not always fit the job and the poor organisation of delivery to farmers meant that they were not always available when needed. The workers would not work at the appropriate time and this had a negative effect on the timeliness of cutting and quality was lost. The drivers of the balers lacked the expertise to drive them, the baler twine had not been ordered and some one had to be found actually to make some. The fields where farmers had themselves prepared the ground were too stony, too rough (due to deep ploughing), and too eroded to take the machines that were available (JDFP, 1985, 2 (1 & 2). pp. 4 & 5).

(b) Straw baling

The project switched its attention to straw baling and in the first season made 50 000 bales. It was said to be the first time this had been attempted in Kerak and Irbid. But it seems to have been no more adopted by the project farmers than the haymaking had been. The aspiration of the team for stone-free fields was reinforced when a stone-roller was imported, but at the end of phase two no farmer had been persuaded to use it (JDFP, 1984, 1 (15), p. 5).

In 1985 the team decided to eliminate the order for a stone-picker from the equipment list and to substitute a small seeds thresher and a pasture meter to be used by the research program (JDFP, 1985, 2 (3), p. 8). In 1986 the team reported that 'no fodder conservation' had taken place, although a drought contributed to this decision (JDFP, 1986, 2 (6), App. 3).

Shallow cultivation

In Algeria there were no scarifiers and no combine seeders, and no change was made to the commonly used deep ploughing.

In Libya each farmer was provided with a scarifier and combine seeder and expert medic farmers showed the farmer how to use and maintain them to prepare seed beds and sow their crops. Shallow cultivation became the norm.

In Jordan the project began with two scarifiers, two combine seeders, one set of heavy harrows, one twentyfour disc offset cultivator, a contra-rotating disc mower, a roll-a-bar hay rake, two hay balers, one stone-picker, two boom sprays and one 1000 litre tank. The World Food Program (working also in Jordan) made available to the team 2/65 h.p. tractors and two hay-balers.

The machinery expert criticised some of the implements acquired. He had no time for disc-ploughs and claimed that their use was in part responsible for the widespread erosion and land degradation in Jordan. He blamed the cupidity of local dealers for their widespread adoption after the 1950s but seemed unaware of the degree of persuasion that occurred through aid programs from Europe and America that included deep ploughing implements as one of their major components, reinforced by machinery manufacturers eager to acquire markets by providing training in their operation and maintenance.

He agreed with Saad Shamout of ICARDA who characterised the introduction of mechanisation into Jordan in the 1950s as 'a curse rather than a blessing, since neither appropriate machinery nor its proper utilisation were considered' (JDFP, 1984, 1 (15), p. 4). He saw to it that one of the two disc-ploughs ordered by the management was replaced by another scarifier.

The wide-line harrows with wings ordered from Australia cultivated efficiently but were not suitable for working small farms on which the project operated, and, because of the need to move often from one site to another, were 'extremely dangerous when negotiating hilly and twisting roads' (JDFP, 1984, 1 (15), p. 16). Later, during phase two, in spite of these warnings, another set was ordered (JDFP, 1985, 2 (3), p. 9).

The haymaking equipment was damaged when used on farms where the project or the JCO did not carry out the cultivation program using scarifiers. The roughness of the ground after the passage of disc or deep plough reduced the efficiency of the mower blades, and stones caused constant breakages.

Other pieces of machinery and equipment were bought by the project over time. The livestock officer ordered some mobile sheep yards and weighing equipment from Australia for a grazing trial. A further two four-wheeled drive pickups were also bought (JDFP, 1985, 2 (1–6)). In the final phase the project imported in a vacuum seed harvester for a seed production program.

The scarifiers and combine seeders did have an effect, however.

The attack on deep ploughing

The project in Jordan took a different approach to the attack on deep ploughing mounted in Libya. Although deep ploughing on Jordanian farms was not as deep as in Algeria, for example, it was regarded as necessary for moisture conservation and proper working of the soil. Farmers who had their own implements used discs and chisels and sometimes a mouldboard plough to prepare their ground. One saw the usual long, difficult cultivation program being performed leaving deep ruts and large clods behind.

Although the plan was to work through the JCO to introduce shallow cultivation, there were some initial attempts to use the farmer's own machinery to carry it out but this proved 'unsuccessful' and the team decided to limit their activities to those who contracted the JCO to cultivate their farms.

This may have been unfortunate as it had the effect of somewhat distancing the project team from its original intention of working closely with individual farmers, but the fact that the farmer's implements did not adapt to shallow cultivation was probably a big factor in the decision.

The project's scarifiers and combine seeders were used by the machinery expert to carry out cultivation on farms where the medic/vetch forage mix was distributed and sown, and these implements were also used by JCO tractor drivers (under the supervision of the machinery expert) on farms where the JCO was contracted to prepare the seed bed for the cereal crop.

The demonstration provided by the machinery expert rapidly converted the JCO to shallow cultivation because it was cheaper and quicker than deep ploughing, and when aid funds became available for machinery purchases they bought more scarifiers and combine seeders and began to use shallow cultivation and machine seeding for all their contract work.

The machinery expert was able to train operators to carry on after he left, and he persuaded the GTZ agency that scarifiers for shallow tillage and combine seeders should be recommended for the machinery centre they were planning.

The machinery expert went directly into the farming environment with its endless variables and met them head on. He carried out shallow tillage on real farms and the results spoke for themselves. No trials were carried out into the difference of yield obtained although one was later suggested.

In Jordan shallow cultivation was adopted as an improvement on its own account, not simply as a necessary component to a medic system.

Regeneration of the pasture after the cereal phase

In spite of the rapid adoption of shallow cultivation for the cereal phase, one could not draw a conclusion about its effect on the regeneration of the pasture as very little pure medic was sown on the farms. On the university site, on which shallow cultivation was used, the regeneration was excellent and the productivity of the medic reflected this.

On our visits to Jordan we asked to be shown regenerated medic on farms. Occasionally one saw a field in which sparse medic plants grew. On vacant land and in the Roman ruins we saw medic growing abundantly. The Chairman of the JCO at the time the project started – A.R. El-Nabulsi – was keen and had the team plant some of his large property at Hisban to medic pasture. He was also reputed (on the strength of the grazing he obtained) to have leased large areas of rangeland and established medic there for grazing. At a meeting in 1985 of a newly established project coordinating committee he asked that the project concentrate on trying to introduce medic into dryland areas receiving an average annual rainfall of 200–300 mm.

The JCO wanted some action on the rangeland too. At the same meeting it asked that 50 dunums (5 ha) of medic be sown at Khansari station (annual rainfall below 150 mm) and that medic be sown among atriplex shrubs to see what pasture this could provide. The team leader responded that he was sure that ICARDA would cooperate to help with this suggested program (JDFP, 1985, 2 (1 and 2), pp. 17 and 18).

Inspections of Nabulsi's Hisban site in 1986 and in 1988 showed rather sparse pastures of *M. rotata* and *M. truncatula* (cultivar Jemalong) growing on very stony ground. The pasture was said to have been regenerating for some years and was heavily grazed.

Inspections were also made of a rangeland reclamation site that used medic in the way that the JCO had requested. This is referred to below (Chatterton & Chatterton, 1986 and 1988).

Project objectives

The projects in Libya and Algeria kept to the script. They did their best to demonstrate a medic/cereal rotation and they produced models and guidance for other projects to use and build upon. In Jordan confusion about the objective of the project began early on.

Part way through phase one the team decided to 'test a whole range of alternative management practices and potential forage crops most suitable to the physical and socio-economic environment facing Jordanian farmers' (JDFP, 1981, 1 (1), Annexe C, p. 12).

Subsequent reports record quite a lot of time spent by team members cooperating with other agencies under the umbrella of this broad perspective. The team considered this a valuable contribution to their own project. There were in Jordan at the time GTZ, WFP, USAID, the EEC and the British ODA. All had been in the country for some time and all were well endowed with resources and all were encouraging some aspect of a European or American-style farming system of deep ploughing, the use of nitrogen fertiliser, lot feeding of livestock with grain, and legumes and forage crops as replacements for bare fallow.

In the first season of the project the team spent some of its time advising the WFP on the planting of about 2 ha of mixed oats and beekia for a fodder conservation program. In the final years of the project the livestock expert spent a great deal of his time partnering the British Overseas Development Agency expert in an animal health program and the planning of a brucellosis prevention campaign.

ICARDA took advantage of the team's presence to use them as an agency for the distribution of new varieties of forage crops, which they gave to farmers, somewhat sourly noting that the ICARDA inspired planting of lupins failed in the zone 'as we expected' (JDFP, 1984, 1 (15), p. 9).

The generosity of the World Food Program in making tractors available to the initial team and the cooperation with GTZ that resulted in the importation of scarifiers and combine seeders for the JCO's contractual work with farmers certainly proved valuable in enabling the machinery expert to train many technicians in the operation of shallow cultivation. The benefit of shallow cultivation in ensuring the regeneration of medic pasture after the cereal phase did not achieve the same degree of penetration because there was little if any medic pasture on the farms.

The Australian team had something unique to offer, yet they were only too easily persuaded to copy the same methods used by all the other agencies in the country.

The retreat from the farm to the research site

In phase two the team began to draw further away from the concept of a farmer training project.

The provision of fences to partaking farmers was ended, as was the provision of seed and fertiliser, and subsidies were withdrawn on contract services for seed bed preparation and sowing (JDFP, 1985, 2 (4), p. 3). The annual distribution of medic and various forage crop seeds continued but the relationship between the project and the University of Jordan as well as ICARDA became more dominant.

The team concentrated more on joint research programs, student seminars and involvement in international conferences. The next couple of expert farmers employed as farm technicians were occupied in repairing machinery and carrying out the haymaking and tillage requirements of the project's research plots.

Further 'intense training' of technical staff was planned, but as all the team were fully occupied, there was no one available to undertake it. Two stone-rollers arrived from Australia and were assembled for use on local farms but the 'poor crop conditions' made their demonstration impossible.

A miscellany of interests, dictated by the idiosyncratic choice of team members, took over.

A demonstration of sheep dipping took place and some advice was given to the machinery stations. University undergraduates were given a lecture and assessment program and made a tour of project experiments on Jordanian research stations and saw some farm demonstrations. Later, some machinery operators were shown how to assemble and operate the stone-roller and a farmer and his sons were shown how to put up a fence. Three seminars were given on 'Forages in Jordan', 'Pastures in New Zealand' (a climate similar to England); and 'Weeds and Weed Control in Cereals and Pastures in Jordan'. Some team members gave papers at an FAO-sponsored international conference.

In 1986 the official 'Project Objectives' talked of the team working to improve cereal husbandry 'over 3% of the rainfed area' (JDFP, 1986, 2 (6), Annexe B, p. 15), yet in 1985 the project had declared that the cereal program was being abandoned in order to concentrate on 'agricultural staff and policy makers' (JDFP, 1985, 2 (3), Annexe C, p. 13).

A socio-economic report was produced in a brief interlude in the first months of phase two (JDFP, 1985, 2 (4), p. 7). The team leader reported that 'The study showed that there are few sociological factors inhibiting the adoption of the new technology being introduced by the Project' (JDFP, 1986, 2 (5), p. 2).

The pasture expert filled a quarterly report with detailed plans for experiments and trials for a 'Forage Research Program', most of which were carried out with the cooperation of a new national Centre for Agricultural Research and Technology Transfer and the staff of the University of Jordan on land made available for the purpose.

The team leader reported during this period 'no formal training or extension activities were conducted' although counterparts (reported elsewhere as rarely being able to take part in field work) were given 'extensive on-the job training whenever they were present . . . '. As we have seen above, they seldom were.

The difficulties encountered by the team in getting the JCO contractors to cultivate and sow the research plots at the optimum time was sheeted home to a 'conflict of interest – the farmers pay for their services, the project does not'. The team decided that the farm technician should give first priority to the preparation of the project's trial plots using the farm machinery supplied to the project originally for demonstrations on the farmer's fields (JDFP, 1985, 2 (4)).

In 1989 the Australian aid agency (AIDAB) carried out a review of the project. The review reported that in spite of having 'had a good impact on wheat yields', ' . . . the concept of self regenerating legumes in the rotation' had been 'abandoned' in order to promote a sown annual vetch crop which farmers were slow to adopt. The AIDAB evaluation of the project overall was that 'the [Australian] system is not well suited to local and farming conditions' (Hewson & Stensholt, 1989, pp. 5–6). The fact was that the Australian system had not been tried on farms in Jordan.

The seed production unit

In spite of this discouraging review, a third phase was begun. Another Australian farmer (George Heading) took over as farm manager. The team for this phase was a team leader, a livestock adviser, and Heading as farm manager/technician. The university site continued to show that (given appropriate management) medic pasture would grow in Jordan and produce abundant supplies of green and dry forage for sheep and cattle.

In addition to his duties as tractor driver, Heading developed a sideline of his own – that of demonstrating that it was feasible to produce commercial quantities of medic seed in Jordan. He seems to have persuaded the project management that seed production should be the target for the third phase.

He began by growing vetch seed because the project was encouraging farmers to use it. The technical team preferred the ICARDA-bred Syrian cultivar of vetch but Heading eventually concentrated on the beekia vetch that was commonly grown by farmers in Jordan. In addition he sowed *M. rigidula* and *M. rotata* seed.

Heading used the project scarifier and a combine seeder and so produced the level seed bed needed for good germination and efficient harvesting. As he grew his medic and vetch seed on fenced sites he was able to protect it from unauthorised grazing.

He persuaded the Jordanian authorities to request from Australia (under the terms of the AIDAB contract) a small-seeds harvester and, after it arrived, he harvested 800 kg of *M. rotata No. 2123* and 200 kg of *M. rigidula* medic seed.

In winter 1989/90 he planted more medic seed at three sites and in June 1990 he harvested 500 kg of seed from the driest station (Ramtha) and in the first four hours of work at the best station (Madaba) harvested 200 kg of clean seed off 10 dunums. As a result the project leased 200 dunums (20 ha) for three years just for seed production, but the evacuation of the team due to the Iraq War prevented the conclusion of this particular program (Heading, 1990, private correspondence).

Medic on the rangeland

An unexpected side effect of the project was the result of a gift of some unused medic seed of the Snail variety (cultivar Robinson, *M. scutellata*) to an employee of the JCO who had begun a semi-official rangeland reclamation project in the Ma'in district. He had been one of the counterparts to the project in its early days.

He believed that the conventional closing off of rangeland and exclusion of stock, let alone the use of nursery-raised atriplex plants, was wasteful and unnecessary.

He managed to persuade several tribes with grazing rights over the Ma'in site (average rainfall 120 mm) to cooperate in a management scheme, and with machinery borrowed from the JCO and some World Food Program handouts ripped the site and broadcast seed from local atriplex in addition to the medic seed donated by the project.

Although the cultivar Robinson is not suited to semi-arid conditions and only regenerates effectively at the higher end of the rainfall scale, it did produce pasture and it regenerated. The result was sufficient pasture and atriplex on 1200 ha with which to graze 3000 sheep for 21 days in autumn and 21 days in spring. Previously the site had been bare all year round, but after several years into this program there was good vegetation and some animals and birds had returned. The cost of the reclamation was 20–25 JD/ha compared to the conventional cost of 100–120 JD/ha.

In subsequent years the program was extended, and the tribes paid a small sum as a contribution towards a guardian and the purchase of superphosphate to be broadcast over the site. There were no fences and community pressure and the presence of the guardian seemed sufficient to prevent unauthorised incursions. The Robinson medic did not flourish but the management of the site was such that there was a good re-establishment of the spontaneously occurring local medic and the improved quantity of feed available has continued (Chatterton, 1988; Hesheiwat & Mahommad, 1987).

This project was used as one of a series of baseline studies for a US $20 million FAO/UNDP project between 1984 and 1988 (*Rangeland Management*, 1984–9) and is referred to in a workshop on pastoral communities held in Jordan in 1991 where it was reported that access to the pasture and shrubs has been broadened to include nearby farmers (FAO, 1991, p. 8).

The problem discussed at the site in 1987 seems unresolved; that in summer, livestock quickly exhaust the sustainable levels of production from the atriplex and must be withdrawn to protect the shrubs before the available supplies of medic pods and straw are utilised. It is not clear if any further sowing of medic seed has taken place.

A review of the project

Attempting to introduce a new farming system into a community where there are entrenched farming practices reinforced by local technicians clinging to the old system is very difficult. A medic/cereal rotation can be established relatively rapidly, but in dryland conditions it takes a number of seasons with variations in rainfall and other climatic factors to demonstrate its reliability to farmers. A longer term provides the opportunity to demonstrate the inherent management flexibility of the system, to illustrate its cheapness *vis-à-vis* the existing system, to discard or purchase new and more appropriate machinery, and to modify or expand project directives. It enables a longer term assessment of the performance of various cultivars of medic under farming conditions. It provided teams with the possibility of building up strong relationships with local technicians and farmers and developing a team approach to extension programs.

The Libyan projects ran for nearly ten years and demonstrated not only that the system was sustainable within a typical range of seasonal variations met with on dryland farms, but also that Libyan farmers could manage the system themselves given the correct implements and advice.

Because the models used to demonstrate the phases of the system in Libya were clear and the Seedco farmers provided the information needed by the farmers, the result was a landscape of farms on which there was a recognisable medic/cereal rotation.

The Jordan project ran for eleven years, but in 1988 when filming shallow cultivation in Jordan for the FAO audio-visual kits it was difficult to find a scarifier still fitted with a depth wheel although wide-line scarifiers were being used by the JCO contractors to cultivate large private farms. The JCO contractors were sowing crops with combine seeders, but many smaller farmers were still putting their fertiliser out by hand.

The JCO and the Ministry technicians could not produce one farm with a recognisable medic pasture being used either alone or in rotation with cereals, but dry medic pods were on the ground in abundance on the university plot used by the project team and these were the surplus remaining after a period of controlled, heavy grazing.

Apart from further proof that dryland farmers find haymaking too expensive to adopt, the project in Jordan did show that shallow cultivation had advantages over deep ploughing apart from its effect on medic regeneration. It also showed that if cultivation and seed bed preparation techniques were improved a local medic seed industry was not impossible.

Heading's seed production program succeeded but it did not become a model to be used, for example, in Morocco when seed production was attempted, nor when ICARDA was tortuously pursuing its own program to try to develop seed production units in the region. ICARDA blamed the harvester for their difficulties, not the seedbed preparation techniques.

This illustrates the futility of isolated successes unless they are integrated into a continuing and coherent program. If this seed production success remains isolated from the mainstream of attempts to establish medic in the region what is its benefit? Once Heading left the project who would continue the program? Who would drive the tractor and seed harvester and make the decisions about varying the program in response to seasonal variations in climate? Unlike the machinery expert, Heading had not been in a position to train operators and managers to succeed him.

Will the medic seed he harvested be efficiently used if it is sown in badly cultivated, rough seed beds by farmers who cannot afford, or do not choose, to have the JCO come in and prepare their seed beds? Will they lose it if it fails to regenerate because it is ploughed in, or lose the seed bank if it is overgrazed during flowering time? These questions indicate the degree to which the Jordan project failed.

One can say that it is ridiculous to expect a small team to come into a country like Jordan and change the face of its dryland farming, but one could have expected that after eleven years with a reasonably large budget and the opportunity of using the appropriate machinery and carefully selected medic seed, a nucleus of farmers would have been shown how to establish a medic/cereal rotation, if not perfectly, then at least sufficiently to indicate the benefits of it to other farmers.

Over the same period, the Seedco farmers in Libya taught 600 farmers how to manage the system.

Reasons for failure

The teams considered they failed to demonstrate a medic/cereal rotation because of

- a lack of understanding of the management of a ley farming system,
- the complexities involved in growing crops and keeping livestock on the one farm,
- the lack of an extension service familiar with the requirements of ley farming,
- the 'newness' of medics,
- the lack of government support for livestock production and markets for hay and grazing rights,
- and, of course, poor nodulation of medics – a problem that they later admitted disappears in the second year (ICARDA, 1989, p. 76).

But there were other reasons.

(a) Management

Although the first phase was for five years, the project operated only for three and a half years. In the first season the team had difficulty in organising a program. The next two seasons were affected by drought, one gravely. In the fourth season the project was inactive because of the hesitancy of AIDAB in making any decision about whether the project would, in fact, continue. The fifth season was the handover period to the phase two team and no active work was carried out. The third and final phase was interrupted by the Gulf War.

The intention to undertake tests of a whole range of alternative management practices and forage crops was not only unrealistic, but presented a profile suspiciously like the conventional project from the better-funded European and US donors. The conventional forage crop demonstration could surely be left to teams working for the World Food Program, GTZ and USAID, all of whom had much greater access to resources than had the Australians. These agencies had, in one form or another, concentrated on forage crops for decades and not succeeded very well and there was little point in emulating them. Time spent on joining the chase for these hares took up the time and resources of a team who for once had something different to contribute.

Had the management from Australia been more interventionist and clearer about the fundamentals of a medic farming system, this may not have happened.

(b) Research

The tendency of the technical experts to concentrate on research rather than to tackle the practical tasks requiring attention on existing farms took up so much time. Three experiments (in both wet and dry years) were carried out to confirm that the best time to seed cereals and forage is between mid-November and mid-December. Every farmer worth the name knows that the best time to sow is immediately after opening rains, and every Jordanian farmer knows that, in Jordan, the opening season for cereals is between mid-November and mid-December. Yet the team took time to report that the Ministry of Agriculture in Jordan has a policy of advising farmers to sow as soon as possible after the opening rains, and that the results of the project trials confirmed that 'autumn rains are extremely variable in Jordan, and produce between 10–15% of annual rainfall and range between 0–30% in a given year. A "good rain" is considered to be 15 mm., and most farmers sow their cereals after one or two good rains' (JDFP, 1984, 1 (15), p. 7). The irrelevance of this work to the farmers who obviously needed help to lift their soil fertility and feed their sheep better is stark.

(c) Influence on farmers

One of the reasons given by the team for failure is the lack of an effectively trained extension service in Jordan. In spite of the decision of the team half way through the project to give up on the scheme to train counterparts from the JCO technical staff, to abandon advisory work with farmers, and to select graduate students to extend an understanding of medic, there is still not in Jordan a recognisable group (as there was to be in Iraq, Tunisia, Algeria and Libya) in the middle and top levels of agricultural administration who are prepared to make policies and persevere with the challenge of getting a medic system established.

There is no group of farmers who can be said to have successfully established a permanent medic/cereal rotation, and no core of local technical advisers who have sufficient experience of its seasonal requirements to effectively advise farmers who wish to adopt it.

Conclusion

By 1979 when the Jordan project began, the Libyan projects were well-established. The South Australian agency was manager for both the Jordan project and the Jabel el Akhdar demonstration farm. Yet there was no apparent correspondence between the two teams. The models developed at Jabel el Akhdar and the Gefara Plains, reinforced by the experience gained during the FAO project in Algeria were, with the exception of the introduction of shallow cultivation, not used.

The social and economic conditions and climate of Jordan were not so vastly different that these models were irrelevant.

- In the case of shallow cultivation, the modification to the program undertaken in Libya was that a contractor, rather than individual farmers, was supplied with implements and training. The result was the same – there was a change to shallow cultivation.
- The cultivars supplied in Jordan were the same as in Algeria and Libya and the only difficulty with them in Jordan appears to be when they were sown in trial plots that were not managed well. There was no report of them not growing satisfactorily in forage mixes or on the El-Nabulsi farm or, even on the Ma'in rangeland.
- Farmers in Jordan with strips of land did pose problems when it came to having a rotation of cereal and pasture, but the project soon moved its focus to farmers who had holdings of 10 ha or more in one piece and as they accepted a forage mix one can assume that they had a small flock of sheep which they ran in conjunction with their cereal crop. Farmers in Morocco and Tunisia with similar landholdings and flock sizes later proved that medic introduced onto small farms was both beneficial and profitable.
- The absence of expert farmers who knew the system and could adapt it on the existing farms was a serious liability. None of the difficulties presented by the technical team were beyond the capacity of expert farmers with experience in the region to overcome. Had they been involved, the original intention of the project to establish a medic/cereal rotation on Jordanian farms would, in all probability, have been achieved.

8
Two medic projects in Iraq

Introduction

In the mid-1970s the Iraq Government also became interested in the results being obtained in Libya. An official delegation went from South Australia to Iraq in 1979 and Iraqi Ministry officials were invited to Libya to look at the El Marj and Gefara Plains projects. They continued their journey to Australia to discuss technical cooperation and the possibility of similar dryland farming projects being carried out in Northern Iraq.

Dryland farming in Iraq

The cereal zone of northern Iraq is similar to much of the cereal zone of Jordan, but instead of turning to dust when it is exhausted, the soil settles down into a sulky clay that caps after rain and prevents seed from emerging. This extremely poor soil structure is the result of years of overcropping and erosion and a yield of 600 kg/ha of cereal is regarded as good. Pasture is scarce and sheep depend on tibben (a mixture of chaff and grain) and concentrates for most of their nourishment (APDP, 1982, p. 21). During an inspection in 1979 the cereal farms appeared in much worse condition than similar farms in the Jebel el Akhdar and the Gefara Plains in Libya.

The farming system

The farming system used for cereal production is primitive in spite of widespread mechanisation – deep ploughing, more working of the soil to prepare a rough seed bed, fertiliser and seed broadcast on the surface, and a rotation of wheat after wheat with occasionally a bare fallow (APDP, 1982, p. 44). Large flocks of sheep are owned by nomadic tribes but most sedentary farmers also have a small flock to provide meat for the family and some income from lambs. Purchased grain, straw and tibben are supplemented by standing cereal residues. There is rarely any pasture for grazing.

The projects

The Iraqi Government decided to follow the Libyan example of setting up a demonstration farm in a relatively fertile region to provide an illustration of the medic system in full operation, as well as several large scale development sites on which the system would be established in the more marginal zone. Both projects were to include applied research to ensure that medic cultivars and fertiliser applications, for example, were suitable for Iraqi conditions.

What was equally important to the Iraqi Government was their wish that a strong extension component be undertaken by both projects as soon as possible in order to enable their farmers to benefit from this new system.

The South Australian agency (SAGRIC International) was awarded the demonstration farm project at Erbil and the Western Australian Overseas Project Authority took the contract for the development sites around Telafar. Each project had a term of five years.

Once the initial work had been done to establish the suitability of medic to the two sites, the Iraqi Government intended to create individual dryland farms of about 100 ha on which farmer families would keep sheep and grow cereals. At the same time the existing private farms (which were owned by inhabitants of the mud built villages in the vicinity) were to benefit from the new techniques by being involved in demonstration and training programs. There is an agricultural university at Mosul and a cereal research centre near Erbil and it was hoped that these institutions would work closely with the Australian teams (Chatterton, 1979, 1980–1).

The Western Australian project

The contract signed in 1979 resulted in medic and cereals being established on seven sites on the Upper Jezirah District of the Upper Plains and Foothills region of north-western Iraq. The sites were at Telafar, Rabiaa, Hammam Al-Alil, Hatra, Ajnadine, Baaj and J'Ravi. Rainfall recorded during the growing seasons on the sites during the project varied from less than 100 mm at J'Ravi (average about 250 mm) to 441.3 mm at Rabiaa (DDAJP, 1985, p. 13). The team established 7000 ha of medic at J'Ravi, and 2500 ha of medic/cereal at Ajnadine, together with about 125 ha of medic on the experimental centre at Telafar that was used to provide a demonstration of the pasture in rotation with cereals.

The performance of medic pastures

All sowings were on project sites with the exception of some communal land where a grazing trial took place. Australian cultivars, Borung (*M. truncatula*) and Serena (*M. polymorpha*), together with some of Circle Valley (*M. polymorpha*) when seed was available, were used for the establishment sowings. A program to select, multiply, and test local ecotypes and to select rhizobia strains compatible with them was 'highly successful' and at the end of the project a range was available for large scale testing (DDAJP, 1985, p. 16).

It is difficult to assess the performance of the medic overall. Certainly large areas of medic were seen during January 1983 to have germinated well and were growing in spite of the severe cold. The community grazing scheme did not report any disappointment with the amount of grazing provided by the newly introduced medic pasture. It was noted that the Australian cultivars Borung, Circle Valley and Serena (*M. polymorpha*) although comprising only about 30% of the 1980 planting, provided about 80% of the total seed yield by the end of the second year (DDAJP, 1985, p. 18). Photographs taken (some of which appear in the *Final Report* prepared by the Western Australian team) also show medic growing strongly at various periods during the year. However, the team reported that 'failures occurred on extensive areas' (DDAJP, 1985, p. 17), and this was thought to be due in part to the extremely low fertility of the soil and an absence of the bacteria that is required for nodulation, in part due to the poor structure of the soil leading to heavy capping after rain that prevented medic seedlings from emerging, and in part to the unexpected problem of wholesale bird-grazing that occurred and at times swept the ground of medic seedlings.

It was observed that some cultivars (Circle Valley and Borung) proved relatively tolerant to this grazing, although they were attractive to rodents (*Meriones* spp. locally known as 'Jerds') which invaded the project site (DDAJP, 1985, p. 17).

The climate and conditions during the term of the project were severe. Average daily temperatures in July and August were at times over 40° C and during January and February were less than 5° C, while grass temperatures (in January 1983) fell as low as −15° C on one site.

The growing season extended from October/November through to May, although recorded growth was slow until March due to cold, after which growth was abundant. Some introduced cultivars of medic could not stand the cold; others survived well. The local ecotypes (found in abundance on roadsides and wasteland) were on the whole much better at coping with the climate. It was noted that the maturation and pod set characteristics of these local ecotypes were 'best fitted to the normal prevailing temperature conditions for pod set' (DDAJP, 1985, pp. 17 & 18).

There was a shortage in the supply of medic seed available to the project. The Australian cultivars (Circle Valley and Borung) were not available in sufficient quantities, nor was Robinson (*M. scutellata*), a commonly produced Australian variety. Seed from the various Iraqi ecotypes identified and named Cl 9.1 and 9.2 etc., and Cl 43.7 (which 'retained its promise throughout the Project'), as well as Cl 12.5 and 14.2 (ecotypes of *M. rotata* identified and named in Syria at ICARDA) were the product of experimental plantings only and commercial quantities of this seed were not available. These cultivars in the experimental plots seemed not to be attractive to the predatory birds and had a definite resistance to frost. There were obvious advantages to be gained from using the local ecotypes.

One problem with them was that most of them were spiny – 'as is indeed the case in natural selection of medics around the world' and it was feared that this might be a disadvantage to farmers who wanted to produce wool. As a brevispina variety of *M. polymorpha* appeared to be rare in the region the team felt that in selecting local cultivars for commercial production it would be 'probably best to eliminate only the types which will give the greatest contamination in wool'.

The option of encouraging hand collection of pods of existing local medics by farmers to enable them to extend their initial pastures was not envisaged.

The effect of seasonal variations on medic

Data on seasonal variations recorded for the project zone shows that 'Many years are failures for cropping. South of Sinjar there is one crop failure in three years, north of it one in five years' (Buringh, 1960; DDAJP, 1985, p. 12). This is very similar to districts in South and Western Australia with similar rainfalls.

Although 1983/4 was the driest year on record the team reported that 'It was heartening to note the successful regeneration of medic at Ajnadine in the 1984/5 winter *following a crop and then a year of drought*' (DDJAP, 1985, p. 17).

Owing to the effect of severe seasonal conditions on the productivity of the medic, the results varied from 'disheartening' to 'cause for optimism'. Particularly notable was the 'good establishment on 2,500 ha. at J'Ravi in the adverse 1982 season' and the successful regeneration of medic at Ajnadine in 1984/5 'following a crop and a year of severe drought'.

Yields of medic seed at Ajnadine (average rainfall during the term of the project was 349 mm) were 250 kg of pure seed from cultivar Circle Valley and 150 kg from cultivar Serena in 1981 following a cold, dry season and significant bird grazing (DDJAP, 1985, p. 53).

During the project, trials undertaken during the best season with local ecotypes of medic (9.1 and 9.2) produced 2800 and 3000 kg dry matter/ha, and it was deduced from this that even in an 'average season' production of 1500 kg/ha of dry matter from local ecotypes would (if medic were used to replace the current fallow phase) provide sufficient feed for some four million sheep in the Jezirah area alone (DDJAP, 1985, p. 20).

Grazing trials

It appears that the land for the project sites was simply resumed by the Iraqi Government from local farmers, although at J'Ravi the pastures sown on what was communal rangeland were grazed in cooperation with the flockowners living in adjacent villages who had grazing rights over the land. The final report refers to 'grazing throughout the project being carried out by local sheep with various owners' and flocks being used for this purpose were seen during an inspection of the project (DDJAP, 1985, p. 6). At J'Ravi, which was the driest site, 2500 ha medic in a particularly cold and dry year produced 55–83 kg/ha of seed and the stocking rate varied from 2.5 to 4 sheep/ha with supplementary hand feeding becoming necessary from early November to early spring. It was estimated that a year-round stocking rate of 1.5 sheep/ha with a little supplementary feeding now and then could be reasonably expected. The team considered that a grazing regime of less than 1.5 sheep/ha would not satisfactorily control weed growth.

Flocks from local villages were brought in to graze the pastures established on some project sites but no detailed account is given in the reports of the organisation of this grazing, nor of what agreements were reached with local flockowners. The Project Director noted in the final report that 'unfortunately the lack of available sheep to keep close grazing of the medic pastures prevented a proper evaluation of the medic pasture–cereal rotation ... ' (DDJAP, 1985, p. 7).

The possibility of unauthorised grazing by nomadic flocks was dealt with in the following way. When fencing the sites, passageways were left open for transhumant flocks to move from one traditional grazing site to another – a cautious acknowledgement of the 'nomads cut fences' threat, but an advance on the earlier belief that nomads should be kept away altogether. The sites were fenced externally and subdivided (150 km of fencing in all) and when the project finished in 1984, the fences were carefully removed and returned to the project base and the land of the J'Ravi site and the Ajnadine farm was returned to the local farmers (DDJAP, 1985, p. 53). The Director was disappointed that the 'large scale fencing and big areas of medic planting will not be of direct value because the areas have reverted to the local farming systems' (DDJAP, 1985, p. 7).

Buildings, plant and equipment

A four-stand shearing shed was erected, although the only shearing that took place was a demonstration by (admitted) amateur shearers among the project staff. The shearing shed was built with a floor of fully imported (and very expensive) Western Australian jarrah wood that would not have disgraced an elegant ballroom. Elaborate watering systems and large scale farm machinery were imported and installed. A swimming pool was built in the Western Australian compound and houses and a recreation centre were comfortably equipped. A video recorder was bought for making extension films. Scarifiers, combine seeders, tractors and other implements and equipment for scientific measurement were imported.

Shallow cultivation

Local farmers used discs and ploughs to prepare their cereal seed beds and it was acknowledged that this must change if a successful medic/cereal rotation was to be adopted. Groups of Western Australian farmers were flown in each year to drive tractors and use scarifiers and combine seeders to sow the extensive areas of medic and cereal. The continued reliance on Australian farmers to perform the sowing program led to complaints by the Iraqis that no local Iraqi farmers were being trained in the use of the equipment used for shallow cultivation, or in the operations of seeding, fertiliser application and so on (Chatterton, 1983). When in response to this some attempt was made to include Iraqi nationals in this operation, there was reluctance to admit the ability of the Iraqis to acquire the necessary expertise and it was found more convenient to continue the trouble-free and less time-consuming strategy of importing the Australian farmers.

Contact with local farmers

The *Final Report* leaves an impression that the idea of involving local farmers in the operation of a medic/cereal rotation was not a top priority with the project management. Although Iraqi counterparts were appointed to the project, they were regarded as technicians to be trained in research rather than in farm management.

The final recommendations are indicative of the distance between the expressed need of the Iraqi client for farmer involvement and the inclination of the Australian management.

The four major programs recommended as the basis for continuing work by the team were all designed to continue and expand the research component of the project. Yet, the conclusion of work done to that time was that from 'a purely technical viewpoint medic/crop rotations . . . are feasible in the Jezira'. This acknowledgement was followed by a warning of the need to interact with 'economic, social and other forces such as land tenure, tradition and grazing rights' (DDJAP, 1985, p. 58).

The *Final Report* contained few signs that these factors had been taken into account during the term of the project. With the exception of the J'Ravi community grazing scheme, local livestock owners, for instance, had been seen as passive providers of sheep to graze project sites, not as significant persons to be consulted about how the medic might fit into their social and economic conditions.

It was recommended that further research be carried out to select cultivars of medic identified for their bird and rat resistance. Yet the team had discovered that solutions to many of their field problems were found by carrying out simple farm operations. For instance, bird and rat grazing could be controlled if the medic pods were buried with a light harrowing; that weeds such as Aran (*Hyperium* spp.) and wild barley grass (*Hordeum spontaneum*) could be kept in a sufficient state of control by mowing or topping when grazing was insufficient or unwise (in the case of Aran which induced photosensitivity before flowering).

These farming solutions became lost in the memory of experience and seldom made any major impact on the list of recommendations for follow-up projects nor were they incorporated into existing Iraqi agricultural extension programs.

The team had demonstrated that medic would establish well and that where the necessary rhizobium was absent the seed would benefit from inoculation with a commercially available rhizobium. They even prepared a film to show Iraqi farmers how easy it was to carry out the inoculation using lime pelletting as farmers do in Australia. They discovered that many indigenous ecotypes of medic already existed that tolerated the climatic excesses of the region, and that cereal varieties when grown in rotation with medic pasture yielded much better, providing weed control was adequate in the pasture phase. Even in a relatively poor season, it had been demonstrated that the supply of dry feed from medic pastures was better than anything else available and they considered that if medic replaced fallow throughout the Jezirah it had the potential to provide the major source of feed for four million sheep.

Extension programs

Could the project have been used as the basis of a farmer extension program?

It did provide a model for the sowing of extensive medic pasture and medic in rotation with cereals.

It demonstrated shallow cultivation and precision seeding.

The team found the time and resources to prepare a film on how to inoculate medic seed with rhizobium using lime pelletting, but it is not known if it was shown to farmers.

The grazing of the project's pastures and cereal stubble by local sheep was not considered by the team to be a success because it led to 'too many grass weeds being allowed to grow and seed in the pastures'. Their solution to this weed problem was a herbicide program – not the education of the surrounding farmers and pastoralists in more efficient grazing management.

In spite of these successes no extension programs were undertaken.

Assessment of the project

The Iraqi Government wanted a transfer of the knowledge of a medic/cereal rotation to their farmers. They were not insensitive to the need for research and the University of Mosul, which was highly regarded as an agricultural institution, was keen to profit from the Australian project. The Western Australians, however, saw the project sites as a means to experiment with medic in a difficult climate and as means of demonstrating machines and equipment from Western Australia that their commercial interests were keen to sell to Iraq.

The priorities are clearly shown in the *Final Report*. The Director was pleased that the project had demonstrated that large scale agricultural machinery techniques using Australian machinery could be successful on large sites, and also that bulk grain handling, Australian-style, was enthusiastically observed by Iraqi farmers. He made it a major recommendation that ' ... that modern seeding machinery and equipment for handling fertiliser and grain in bulk be introduced into the region' (DDJAP, 1985, p. 53).

This entrepreneurial aspect sits uneasily with the Director's insistence that 'The initial proposals [of the Iraqi project] were aimed at carrying out research'. The Iraqis did not agree wholly with this and asked for a more developmental approach so that their farmers could benefit and 'Further development of the proposal during discussions led to modifications to include investigations into dryland cereal growing, cereal–medic rotations and some large scale medic plantings and farming operations' (DDJAP, 1985, p. 6). The Director was not happy with this and believed that 'the decision to include large-scale plantings ... virtually concurrent with the *research aimed at finding out how to do it* ... exposed the Project to high failure risks and some of these did eventuate', although he admitted eventually that ' ... it had the important advantages of identifying a number of problems which may not have arisen in small scale testing' (DDJAP, 1985, p. 6).

He continued to insist that 'The most important part of the Project was always intended to be research work ... ', and referred to the success of trials undertaken to assess fertiliser response, use of superphosphate relative to nitrogen, the importance of weed control and investigations into herbicide use and alternative types of weed control. These were all trials that presumably were being, or could be, carried out in local universities and research centres.

The gap between the scale of operation of the project and those of the local farmers was something of which the project team was aware. The Director was 'disappointed' that the large scale fencing and vast plantings of medic on the project sites proved of little relevance because when the team left the farmers who took back the project sites simply 'reverted to the local farming systems' (DDJAP, 1985, p. 7).

One team member was clear about what he saw as a fundamental failure of the project.

The effective operation of the medic ley farming system does not require farms, buildings and machinery of large size. Many of the problems of the large Project farms would not occur or could be controlled on small farms of say 100 to 200 hectares operated relatively intensively by an Iraqi farmer and his family. On small family operated farms the protection of fencing, and the control of dogs, rodents, Aran and other weeds, and illegal grazing would be simplified with the closer supervision and bigger labour force. The small Iraqi farm could be operated with or without shepherding sheep within the confines of fenced paddocks with or without automatic watering facilities, and with limited building, machinery and equipment facilities (DDJAP, 1985, p. 53).

In the Director's insistence that the project was really to carry out research one can see the fundamental conflict between the Australian technical expert who sees research as the basis of farming change and the clients who see that the change must take place on the farm but cannot persuade the 'expert' to fashion the project to this requirement.

The Iraqi officials complained constantly about the absence of any significant attempt to transfer information about the establishment and use of medic pastures to their farmers during an inspection tour in 1983. They wanted the Western Australian team to be persuaded to change the project to one of extension and training programs for farmers. This was not possible and the team withdrew in March 1985 as scheduled, and no further project was negotiated.

The South Australian demonstration farm in Iraq

The draft proposal initially put before the Iraqi Ministry by the South Australian Government in 1978 was for a demonstration farm with a heavy emphasis on research. The Iraqis expressed displeasure with this during the visit of official South Australian delegation to Iraq in 1979 and it was rapidly re-written as an extension and training project. The Iraqis indicated that this was acceptable to them and it was left to officials of both governments to put together the contract (Chatterton, 1979 & 1980–1).

Site, equipment and buildings

Two sites totalling 5000 ha near Erbil were allocated to the project. The demonstration farm and its compound were constructed on the Ain Kawah site, and a feedlot, shearing shed and storage facilities were erected on the Baharracca site which was six kilometres to the north of Ain Kawah (SAGRIC brief, 1983, Part Two).

The project compound itself was equipped with a large recreation hut complete with television, video recorder and snooker table, and individual houses were provided for team members (APDP, 1982, pp. 90–5). The site was efficiently fenced by an expert imported for the purpose, houses and farm buildings and yards were erected and medic and cereal trials established well within the required establishment period. Video and film equipment was imported for extension and training purposes. Scarifiers and combine seeders were imported for the cultivation and seed bed preparation. Cereal and medic seed was imported for the sowing programs. There was some concern about security for the team when it went out from the compound to work on outlying sites and the Iraqis made available armed escorts, which somewhat slowed down the daily routine (SAGRIC brief, 1983, Part Two).

The demonstration of the system

Cereal crops (Mexipak wheat and Clipper barley) were sown into seed beds prepared with scarifiers. Superphosphate fertiliser was applied. In the first year land which had been continually cropped was used. The yield was 2120 tonnes from 1400 ha. In spite of losses from grazing, fire damage, and invasions of mice, this was a substantial increase on local yields. Better weed control, better seed bed preparation, precision seeding and superphosphate application were considered the reasons for this (APDP, 1982, pp. 7–12). When cereals were sown on land after medic pasture, there was an increase of 1 tonne/ha over and above this (APDP, 1982, pp. 24–8). Some of the barley (125 tonnes) was reaped and stored for feed. A trial of pre-emergent herbicide control of weeds was set up on a small portion of the site.

Large scale sowings were made of medic alone, medic in rotation with cereals, and trials took place of a wide range of medic cultivars for selection purposes. All large scale sowings were a mix of cultivars Snail, Paragosa, Barrel and Cyprus. All medics sown were inoculated using the lime pelletting method but this caused some problems because the lime clogged the seeders. Oversowing at a rate of 10 kg/ha had to be undertaken on about 30 ha to overcome the absence of seed in large areas of the field and the difficulty of germination that occurred due to severe capping of the soil. The extreme cold winter retarded growth. In late December, January and February the ground was frozen for up to 12 hours per day (APDP, 1982, p. 11). In March, with the longer days and more warmth, the pasture grew abundantly. All cultivars regenerated and nodulated well. The new cultivar Circle Valley was particularly successful. This cultivar and Cyprus would ensure the success of medic pastures in these early years (APDP, 1982, p. 26).

Hay was made from a forage of barley and medic and stored for future use, and the opinion was expressed that the key to supporting large numbers of livestock in that particular environment should be fodder conservation. A scheme was proposed to convert medic pasture produced in spring into hay to be used in conjunction with straw and barley grain as the basis of a livestock production unit. One hundred tonnes of hay made with medic/barley and medic/ryegrass was stored in 1982. More would have been made but for 'the extensive areas of medic hay collected [illegally] by the local farmers' (APDP, 1982, p. 21).

The medic pasture regenerated without difficulty after cereals were sown using shallow cultivation but this was only demonstrated on the project farm. Farmers looking over the fence would have seen how efficient the scarifier and combine seeder was compared to the machines they were using (APDP, 1982, pp. 24–48).

Grazing the medic

Attempts by the project team to establish a permanent flock on the farm site failed because it was difficult to get a local flock that was of acceptable standard and to find and keep shepherds that were reliable. Local farmers were invited to bring their livestock in to graze the pastures. An estimated 6000 sheep equivalent (1 sheep = 1 goat or 1/8 cow) grazed the pastures from 'March onwards' in 1982 (APDP, 1982, p. 23). This was productivity over and above the 100 tonnes of hay made and stored. Farmers could see the productivity that the medic promised, but were not able to learn how to grow and manage it themselves. The South Australians dealt with the problems of nomads and their right of passage by creating fenced corridors through the site.

Iraqi farmers and the project

The South Australian project demonstrated that a medic/cereal rotation could be productive even in a cereal zone with a particularly cold climate, but this success failed to be transferred from the project farm to the surrounding farms. This was not due to resistance by local farmers or any refusal of the Iraqi Ministry to consider anything other than research-oriented trials as later claimed by the Australian management.

In 1983, following an inspection and meeting of local officials on the Erbil site during the visit of the South Australian delegation, a suggestion was made that the local farmers be invited to come to the project to discuss their attitudes to the progress taking place. In spite of protests from the team that farmers were too occupied with their own work to visit the farm, within an hour more than 150 local farmers were in the large shed on the project site vocally expressing their opinions.

They were extremely interested in the medic pastures and the rotation with cereals, and were vociferous in their questions and comments. They wanted very much to try the new system themselves and continually asked why they were excluded from what was going on. They were amused at the handling of their sheep by the foreigners, but impressed by the amount of pasture that was available for the livestock to eat.

Later, the President of the Farmers' Union (Erbil branch) continually prefaced his replies to questions about farm size, tractor availability, seasonal operations and so on, with the words 'When are we going to be shown how to operate this system on our land?' (Chatterton, 1983).

The Iraqi Ministry officials accompanying the delegation expressed their concern that the project continued to operate behind fences and that there was no formal attempt to extend the results obtained out to local farms.

The Iraqi Ministry view of the project

In 1982 the team leaders of both South Australian and Western Australian projects were summoned to Baghdad to discuss 'implications' for the future.

The team leaders prepared their 'agronomic results' but were told that the 'technical success of the system had been proved by the Australians and that the next step was to establish ... the 'human' success of the system by showing that Iraqi farmers could make it work on the basis of small Iraqi-sized farms'.

The proposal put before the team leaders was that the Iraqi Ministry would set up twelve small owner/operator farms each of 100–125 ha depending on the rainfall. Each farm would be fenced in accordance with 'the Australian system' and would be equipped with house, sheds, tractor and implements and other farm equipment and the farmer supplied with livestock, fertiliser and seed. In return the farmers would have to agree to manage the farm 'in accordance with Australian methods'. The degree and manner of the subsidisation of the cost was yet to be decided. Half the farms would be given to experienced Iraqi farmers and half to Iraqi agricultural college graduates.

The team leaders did not see how the farmers could be advised adequately about the farm management required, but the Iraqis said that the Australian team members could make about three or four calls to each farm each year and provide guidance.

When reporting this back to the project management, the team leader cautiously recorded that while 'no undertaking was given' about this, he believed that 'it is not inconsistent with the extension component in our contract' (Special report, 1982, p. 4).

The Iraqis asked the South and Western Australians to prepare a report containing both any additional ideas they may have about what would be useful to such a program and some details about the basic farm design, suggested fencing, rotations, sheep numbers, machinery and overall costs.

As the team leader said, 'we could not have refused in the circumstances' and he was encouraged by what he saw as 'the strong vote of confidence in the success of the two Australian dryland farming projects at this relatively early stage'.

He was keen to follow up the proposal because it could lead to trade benefits for South Australia. If the pilot farm scheme could be extended over the 2.3 million ha of arable land in the Northern Cereal Zone, the decision to respond to the Iraqi requests would lead to 'good prospects for South Australian Industry, if not for South Australian consultancy' (Special report, 1982, p. 6).

In 1983 these plans were discussed in more detail with the South Australian delegation but the scheme failed to materialise in the form in which it was envisaged. This was partly due to the financial drain of the Iran–Iraq War, and partly due to the resistance of the Australian management to be taken into a large extension component without a new contract, and eventually led to withdrawal of both teams when their original project contracts finished.

It should have been possible to use the last year of the original contract to meet the Iraqi requests for an extension program. Plans were made to have an extension and training scheme drawn up by the Australian farmer who had been responsible for the Seedco teams at Jabel el Akhdar. Unhappily this proposal was not carried out.

An assessment of the project

There were considerable successes – particularly with the demonstration that medics could grow and produce well even in extremely cold climates. The effect on cereals yields of better seed bed preparation and precision seeding even when grown on exhausted soil and the even higher yields that followed an intervening year of medic pasture could have left no doubt that the medic farming system was better than the one in existence. The productivity of the medic pastures and the amount of conserved fodder that could be harvested when medics were grown in conjunction with cereals must have seemed like a bonanza to neighbouring farmers. Yet, having demonstrated all this, the team were strangely reluctant to proceed to transfer the skills needed to neighbouring farmers.

These Iraqi pastoralists and farmers looked through the fence (sometimes lifting it to take a little of the tempting pasture) and were impressed, but were unable to find the key to unlocking the knowledge for their personal benefit, Dr Ala Adin Daoud Ali (Director General, General Body for Applied Research), who accompanied the South Australian delegation in 1983, used his fluent English to constantly press the case for practical training for his farmers.

Back in South Australia, SAGRIC International continued to insist that the Iraqis wanted more research and that extension was 'for the future'. The proposal for the pilot farms was not responded to by either project agency. A separate contract for the provision of equipment for an irrigation project fell by the wayside as did a proposal that South Australian government provide technical assistance for a grain bulk handling program (SAGRIC brief, 1983, Part Two, pp. 1, 2 and 3).

Conclusion

If one compares the Iraqi projects with the Libyan ones that began a decade earlier one can see that much less was achieved in spite of greater resources.

Without the extension component that linked the demonstrations with the local farms, nothing remained on Iraqi farms after the project teams had left.

One cannot disagree with the desire of the Australian teams to explore a little first, but the exploration was sufficiently advanced by the middle of the term of both projects to enable the teams to produce good medic and increased cereal yields on the sites.

Inoculation and appropriate cultivars dealt with exhausted soil and extreme cold. There seems no reason why the teams could not build on this to instruct local farmers to do the same on their own farms.

Of course, there were no scarifiers and combine seeders available for purchase by individual farmers, but the example of Jordan where a number of scarifiers and combine seeders were contracted out from a machinery centre could have provided a model.

By the 1980s there was a great deal of information available from previous projects to provide guidance for on-farm programs. For instance, it was pretty clear from other projects that to encourage good grazing management it was wise to sow on each farm sufficient pasture for the existing flock and then expand it as the farmer gained confidence and purchased or bred up more sheep. No attempt was made to work with farmers in this way.

Experienced medic farmers were available for extension programs and some were employed on both Iraqi projects, but they were not used to provide expertise to local farmers.

On the South Australian project at Erbil they carried out the tractor work, looked after the machinery and were overseers of farm labour but had little or no say in the general management of the project and little or no contact with adjacent farmers.

On the Western Australian project at Jezirah they were brought in at seeding times and drove the tractors and planted the medic and cereals. They had no direct input into the project management and little or no contact with the local farmers (Chatterton, 1983).

The demonstration farms – whether programmed for research as on the Erbil farm, or as a base for development trials as at Telafar – were a costly investment with no immediate returns for a country like Iraq that needed a rapid end to critical soil erosion and a lift in production of grain and livestock.

The major concentration in each project was on the medic plant, not on the farmer's use of it. On the South Australian project the pull of conserved fodder was strong and grazing was hardly attempted. The farmers' problems remained outside their notice, something for the future.

The provision of seed for an expansion of the results to local farms was not taken into account. The cultivar Circle Valley, which proved to be the best performing medic in cold conditions, was in very short supply as only limited quantities had been multiplied in Australia. The Iraqi ecotypes (the C series) were not available at all in commercial quantities, and the same problems that occurred in other countries in the region – undulating seed beds as a result of fierce deep ploughing, inappropriate seeding techniques, poor weed control, and inefficient harvesting – needed to be changed before a mechanised seed industry could be successfully set up in Iraq.

One could say that the Iraqi institutions could have used the research that did take place as a base from which to mount their own programs, but the failure to involve either counterparts or local farmers in the operation of the complete rotation left a void that could not be crossed once the Australian technical staff had gone. The teams had not demonstrated, even on their research farms under ideal conditions and with the correct equipment and ample labour, the simple farming system perfected by South Australian farmers nor had they been able to compile proof of its cheapness compared to the existing system although a consultant was employed to provide a suggested methodology by which such information could be gathered and assessed (APDP, 1982, pp. 72–6).

What the teams did was valuable – good for publication in scientific journals (and the pity is that most of the work remains unpublished in the region where it is most needed) – but of little assistance to an Iraqi farmer wanting to integrate medic pastures into his own farm or livestock enterprise.

Aftermath

The pre-occupation of the Iraqi Government with war made Australian Governments uneasy and they did not press for further projects. It is not known what happened to the plan to set up the model farms near the Erbil site and to train the farmers to use a medic/cereal rotation. The Kurdish farmers who had looked on at the Australian work at Erbil and wanted to try it on their own farms have probably been dispossessed and are now refugees. Whether the medic on the rangeland near J'Ravi is still providing grazing for nomadic flocks and those of the resident farmers is not known.

Part three

Institutions, agencies, local farmers and technicians

9

Institutions, agencies and medic – 1950–80

Introduction

We have already seen in Part One the difficulty Australian technical and scientific personnel had in accepting the rationality and simplicity of the medic/cereal rotation being operated by South Australian dryland farmers. They believed that fodder conservation was a better way to go than medic pastures in spite of evidence to the contrary. They urged the continued use of bare fallow long after they themselves had seen and measured its destructiveness on farms in the cereal zone. They never really lost their inclination for deep ploughing although the economics and results of shallow cultivation were superior. It was not until 1953 that research centre farms began to emulate their farmer neighbours and gave up the bare fallow/wheat rotation in favour of a medic/cereal rotation.

In Part Two the review of projects carried out to transfer the system to the farms in Near East and North Africa revealed differences of approach between these groups and the effect this had on the degree of adoption by farmers of the new farming system they were being offered. International institutions and agencies are strong on the ground in the region and they spearhead agricultural development principally because they are able to allocate funds and they play a critical role in decisions about the composition and direction of projects and the selection of staff. The Food and Agricultural Organisation (FAO) with its headquarters in Rome was probably the most influential of the international agencies involved in agricultural development in North Africa and the Near East. It was acknowledged as a hub of technical expertise in agriculture and was part of the umbrella of the United Nations which enjoyed great prestige in the mid-twentieth century. The staff for the projects sponsored or supported by the international organisations were mainly freelance consultants formerly employed in departments of agriculture and research institutions. Their agricultural education was derived from technology developed for Northern Europe and they continued to be reinforced in the fundamentals of this through their professional networks. The problems they were employed to solve were remarkably similar to those confronting the Australian farming community at the end of the nineteenth century and the early twentieth century.

When, in the 1950s it was decided to try to replace bare fallow in the region with legumes and later sub-clover and medic pasture, FAO and other institutions and agencies could have been expected to play a critical role in easing the technology involved into the existing system. But did they?

FAO and medic pastures in the 1950s

When the Australian experience of exploiting an annual legume pasture first reached the Food and Agricultural Organisation in Rome it was greeted with interest. It should be emphasised that this was not the integrated medic/cereal rotation being used by South Australian farmers, but the use of sub-clover pastures for livestock production. This transfer of knowledge coincided with the period in South Australia when the research institutions there finally changed the rotation on their own farms from wheat/fallow and occasional fodder crops to the simpler regenerating medic pasture/cereal rotation.

Dr H.C. Trumble from the South Australian Waite Institute who had published a memoir *Blades of Grass* in 1946 about his career and discoveries as a pasture specialist in South Australia had joined the Plant Production section of FAO. Trumble was an enthusiast – both about sub-clover and about grass in lawns – and he took his enthusiasm with him. In 1953, FAO published a handbook entitled *Legumes in Agriculture* (FAO, 1953). As a co-author of the book, Trumble made sure that information about sub-clover pastures (and to a minor extent, medics) and their contribution to livestock production in South Australia was included at all opportunities. The authors claimed that the book was not a scientific text book but one which concentrated on the problems engaging the attention of practical agronomists. The production of cereals and livestock was declining in North Africa and the Near East and the technical fixes of bare fallow and nitrogen fertiliser were not being effective in the semi-arid zone. Fallow was costing farmers a year of production, its moisture retention capacity was doubtful and it was undeniably contributing to soil erosion while livestock were having to depend more and more on purchased fodder or grain to survive. Few farmers used any fertiliser on dryland cereals and none on pasture, and they seemed immune to advice as to its benefits. Deep ploughing was attracting critical attention. It was noted that the rough, cloddy seedbeds prepared using deep ploughing made it difficult to achieve good, even germination of cereal seed. A rotation that provided nutrition for livestock and increased cereal yields yet did not impoverish the soil and was being sought. Trumble's information about sub-clover pastures and their use in Australia for grazing purposes pointed to the possibility of a solution at least to the problem of declining quantities of fodder. The reader therefore can find much data about the relative merits of grazed sub-clover pasture *vis-à-vis* lucerne, grain legumes, and fodder shrubs which were the mainstays of fodder conservation programs. A warning was given that simple cut and weigh assessments of forage and pasture production are not valid means of judging the value of pasture because the complex relationship between animal and pasture and their combined contribution to soil fertility cannot be evaluated in one operation. The nitrogen fixing capacity of sub-clover and to a lesser extent, medic, is detailed and confidently presented. The authors use a nice phrase when describing legumes as 'a nitrogen factory on the farm' (FAO, 1953, p. 2). There is reference to the problem of inadequate rhizobium for nodulation of legume pastures when they are introduced to exhausted soil and a suggestion that applications of superphosphate may well increase the fertility of the soil sufficiently for the rhizobium population to be encouraged to resuscitate thus enabling indigenous legumes to lift production without the need for other intervention. Prior inoculation of newly sown sub-clover seed is described and recommended as another means of overcoming the problem (FAO, 1953, p. 188). The authors (reflecting work that was proving successful in South Australia) suggested that the role of trace elements such as molybdenum, calcium and magnesium in assisting with soil fertility should not be overlooked.

Recommended seeding rates for pasture establishment were very low – rarely more than one quarter of a kilogram per hectare, and it was claimed that pure clover or medic stands were not desirable and could not survive alone. Trumble had strongly advocated pastures of mixed clover and grass in South Australian conditions and believed that both were needed to achieve the maximum benefit from each (Trumble, 1946).

The medic/cereal rotation was only mentioned in passing. The use of sub-clover and medics to supply a green manure for a cereal crop was dismissed by the authors as 'at best inefficient' (FAO, 1953, p. 44).

In the list of genera and species, annual medics were mixed into a subsection with some annual clovers under the 'Non-Hardy group' considered to be of 'some importance in pastures' and referred to as 'medicks' (FAO, 1953, pp. 296–7).

One of the difficulties the technical experts in Australia had in understanding the attraction of the medic/cereal rotation to the farmer was their own insulation from farm costs. Had they been more able to appreciate the value to the farmer of the low-cost aspects of shallow cultivation and a medic/cereal rotation they would have appreciated the utility of the system sooner.

A detailed comparison of the costs associated with the fallow/cereal and proposed alternative rotations was not high on the list of the agronomic problems listed in the handbook, but it was suggested that a pilot farm with resources similar to those available to surrounding farmers be used to demonstrate alternative rotations that might help alleviate the declining yields (FAO, 1953, p. 16).

The enthusiasm with which the grazing potential of sub-clover pastures was presented had an effect and annual legume pasture took its place in the list of alternatives that the FAO set out to explore.

Progress by 1956

In 1956, P.A. Oram was deputed to assess the progress of the search for alternative rotations to replace fallow. He identified a conflict between those who 'were convinced of the value of introducing rotations containing legumes and fodder crops into standard farming practice' in the region and others who 'were not satisfied as to the desirability of replacing the traditional cereal/fallow system particularly in the more difficult climatic zones' (FAO, 1956, p. v).

What was unquestioned, he wrote, was the need to improve livestock productivity in the region, and to aim for better conservation of rangeland and forest, but this depended on the ability of the farmer to produce more fodder from arable land. The predominance of a cereal/fallow system in the region made this impossible. Agreement about the value of legumes of one sort or another as a replacement for fallow was supported by scientific data.

Oram wrote

Research shows that it is possible to maintain or improve cereal yields, reduce costs and provide more animal fodder by growing a legume instead of fallowing. Only in limited regions does this seem technically difficult or undesirable, and suitable species and rotations have been demonstrated to suit most conditions.

Almost immediately a further conflict erupted between those who favoured the introduction of pasture and those who favoured grain or fodder legumes as the replacement for fallow. It was said that 'further studies of the productivity of pastures as compared with arable fodder crops . . . [must be carried out] . . . before recommendations can be made regarding the advisability of including pastures in rotations in the Mediterranean region' (FAO, 1956, p. 46).

While the advocates of grain legumes claimed them as the 'ultimate step in the process of building up fertility and ensuring adequate nutrition for livestock' others said that 'pasture species have some advantages over annual fodder crops – they are usually perennials, or at least self-regenerating annuals, so that annual cultivation becomes unnecessary: they also have great restorative effect on soil fertility' (FAO, 1956, p. 27). It was claimed that sown pastures were rare in the Mediterranean zone and the chief reason for this was that 'although the number of species available is very large, few have until recently been tested experimentally or in farming practice. The difficulties of establishing dryland pastures are greater than with annual fodder legumes and little is known about seed rates, time of seeding, or optimum rates of fertiliser' (FAO, 1956, p. 27).

The differences of opinion that developed about whether grain legumes and forage were a better bet than regenerating pastures evoked memories of the battle between Callaghan with his fodder conservation program at Roseworthy and the South Australian farmers with their grazed and regenerating medic pasture. Most of those who were educated in the Northern Hemisphere favoured a grain legume or a forage crop. The few who knew about the South Australian system favoured a grazed legume pasture.

The tension between conservatives and progressives within the technical community in the face of a need for change also bedevilled the question. The conservatives claimed that 'Adequate information is seriously lacking concerning these zones which are the main source of controversy in the development of rotations in the region'.

The protagonists of change demanded that

Steps should be taken urgently to incorporate rotations including legumes as well as cereals into general farming practice and this will be possible only by speeding up research and short-circuiting anything which is not strictly necessary. At present there is a tendency to follow slavishly all the procedures of testing and comparison at the expense of practical results (FAO, 1956, pp. 7–8).

In spite of the data contained in *Legumes in Agriculture* available since 1953 and the South Australian experience in these matters neither had much effect in Rome.

Thirtythree years later the debate was still going on. In 1989, the progressives were described as those who said 'there is no time for research, the need for extension is urgent, and seeds and machinery must be imported and land sown to medics as quickly as possible'. They were also accused of implying that those who asked for comprehensive basic research being carried out first had 'a vested interest in finding problems, it keeps them employed' (ICARDA, 1989, p. 48). Neither point of view was accurate and this conflict tended to obscure the real issue which was the farmer's capacity to introduce either a legume crop or a legume pasture to his farm in order to resolve his problems.

In 1956 it seemed that there was a lot to do if those who wanted to try pastures as a replacement for fallow were to succeed in convincing their colleagues that it was a sound objective.

From the point of view of the farmer there were other matters to be faced. Whether grain legumes, forage crops or medic pastures were adopted as an alternative to fallow, each alternative proposed would, for its successful operation, need more than the acquiescence of the farmer in accepting seed and advice. Each alternative proposal required the farmer to practise a different system and to buy different and often additional machinery.

- Grain legumes required more machinery to plant what was in effect another crop at the same time as the existing cereal crop and also required harvesting and storing.
- Forage crops also required more machinery in order to plant the additional crop. Not only had the crop to be sown in competition with cereals but haymaking machinery was essential. The management of feeding the flock became a matter of accumulating additional feed to use while the forage crop was being grown.
- Grazed pastures needed shallow cultivation of the cereal seed bed to ensure the regeneration of the pasture, and a good balance between the number of sheep in the flock and the amount of pasture available.

Although these matters were identified by technical experts they seemed not to place too much importance on them during the following years.

To plough deep or shallow?

Although deep ploughing was being criticised by many technical experts within the institutions in the region, the question of its validity in semi-arid conditions continued to be argued and no changes were made to the type of farm machinery being supplied.

The Oram handbook referred to the large clods and uneven surface of the seed bed in Mediterranean countries as a result of deep ploughing and the negative effect on the germination of seed. Data were available to confirm that depth of ploughing was not a strong determinant of crop yield in dryland zones. One trial quoted in the handbook took place in Libya in the El Marj region. Yields following the use of a wooden plough were contrasted with those following ploughing with four mechanised models, discs and plough, but not a scarifier. The highest yield occurred after the use of a disk harrow where no inversion of the soil took place, and fertiliser proved least beneficial where the plots were cultivated most deeply (FAO, 1956, Table 2, p. 14).

Other technical experts drew attention to the excessive amount of machinery needed to prepare a seed bed when deep ploughing was used and particularly when the long, bare fallow required several cultivations. The incorporation of a legume crop in place of the fallow would only exacerbate the problem. How could farmers respond to government exhortations to grow more wheat and attempt to sow a legume crop when both crops competed for their time and machinery simultaneously? It was pointed out that such conflict could only be avoided if 'intensive mechanisation can be applied to achieve rapid cultivation in the short time available after the rains had begun'. The matter was left open as no solution seemed available. No one seemed to know that shallow cultivation using scarifiers and combine seeders had long ago in South Australia provided just the 'rapid cultivation' needed.

Grain legumes and forage crops

Those who in 1956 favoured the introduction of a grain legume or forage crop as a replacement for fallow were triumphant, even though they appeared to either ignore, or be ignorant of, the difficulties that this would cause for farmers. They were familiar with the actuality of grain legume and forage crops in rotation with cereals on European farms and to them it seemed the obvious solution.

Demonstrations of this rotation, they said, could be made on State or pilot farms and 'rotation experiments involving the comparison of several different systems in terms of effects on soil moisture and fertility as well as on crop yields could probably be dispensed with'. In the more problematical zones, specific research 'should be parallel to the direct introduction into the demonstration phase of simple improvements, such as the substitution of cereal/forage for cereal/fallow'. There was, after all, 'ample experimental evidence in favor of substituting any satisfactory annual legume for fallow in rotation with cereals' (FAO, 1956, p. 10). Already some farms in the region and many of the farms on national research stations were sowing crops such as Berseem, vicia (local vetch), peas, beans and lentils in rotation with cereals and lucerne was a much favoured forage crop even though it required irrigation and could not be grown in a regular rotation with cereals.

None of the trials included medic or sub-clover pasture, although the 'adaptability' of medics to lower rainfall brackets was agreed. The problem was that 'Research in pasture development is probably more complicated than with the simpler type of rotation, since the value of the pasture is difficult to assess, and management – about which little is yet known – can affect this greatly'.

Some progress in dispelling this fear of the unknown seemed likely when it was proposed that 'special studies should be initiated under FAO guidance to establish the relationship between pasture production and livestock husbandry and to disseminate information on methods of range and pasture inventory' (FAO, 1956, p. 28).

In spite of the strong possibility that legume pasture was able in semi-arid conditions to prevent soil erosion, improve soil fertility, increase yields and provide cheap and adequate grazing for livestock – in other words, solve the problems the technical experts were faced with – the decision was that 'In view of the urgency of a solution of the alarming fodder situation, it seems undesirable that research or extension work on the inclusion of pastures in rotations should be given priority over the more straightforward solution of the inclusion of leguminous fodder crops' (FAO, 1956, p. 31). The absence of any real expertise within the institutions about the way medic pastures could stabilise and enrich a dryland farming rotation put paid to further investigation for the time being.

This lack of experience with a medic/cereal rotation among the technical experts employed by FAO and other institutions involved in dryland farming development in 1956 was foreshadowing a future problem. Some of those who only knew the general theory of a medic farming system were already moving into the role of 'experts' within the fund-distributing institutions and their lack of actual farm experience with the system became obvious when projects were planned to include medic pastures as a replacement for fallow. They did not understand the need to provide scarifiers for shallow tillage, for instance, and they tended to regard the medic pasture as a forage crop rather than a pasture, thus allowing it to be grazed in an unsatisfactory manner. In addition, the terminology surrounding legumes became confused. Generally, when legumes were being discussed in relation to dryland farming a grain legume crop or lucerne (*M. sativa*) was meant. Legume pasture tended to remain something of a curiosity and when it was discussed, sub-clover was usually meant.

Farmers' resistance to grain legumes and hay

The decision by FAO in 1956 to continue to give top priority to legume crops rather than to undertake an investigation into the potential of pasture was curious in view of the fact that there were already signs that advice to sow legume crops to replace fallow was not going to be adopted readily by farmers. One reason was the insistence of governments that more wheat be sown. Another was the cost associated with the extra cultivation and sowing program required to produce a crop of legumes as feed for livestock. In addition, many farmers could not find or afford the machinery with which to harvest legume crops. If they were induced to make hay they were faced with the cost of haymaking machinery. Small farmers in particular were unable to replace their fallow with legume crops because they did not have the time or the machinery available to sow what was, in effect, two crops in the one season.

On the other hand, the medic/cereal rotation dealt with all three problems – it produced more wheat and livestock feed than the existing system, shallow cultivation enabled seed bed preparation to be carried out more rapidly and cheaply than with deep ploughing, and the regeneration of the pasture in subsequent years halved the cultivation program. The pasture did not need harvesting or storing. These benefits applied equally to small and big farms but they were unknown to the technical staff who were making the decisions.

It is notable that FAO, which prided itself on being in the forefront of technical farming knowledge, had so little appreciation of the dryland farming revolution that had swept Australian farms. Visiting technical and scientific experts carried out study tours and Australia was no more remote from the rest of the world than parts of Africa or India or South America. Given the critical nature of the falling productivity on farms in the Mediterranean zone, and particularly in North Africa and the Near East, one would have expected the enquiring scientific mind to immediately grasp the significance of the dryland farming rotation that had dealt with the same problems in Southern Australia.

Medic in the 1960s

Little was heard of medics in the 1960s in the Northern Hemisphere although this was a time of great consolidation of the medic farming system on dryland farms in South Australia, especially in the marginal zone where the cultivar Harbinger proved particularly suited to farms where wheat and sheep production had to survive low rainfall and frequent drought. Technical and scientific papers in Australia abounded giving details of enthusiastic seed collection and selection programs and the prodigious nitrogen fixing capacity and productivity levels achieved in an integrated cereal/livestock system using medic pastures, but few filtered through to the Northern Hemisphere.

Medic and the institutions and agencies in the 1970s

By 1970 some of these technical papers from Australia must have made the journey to the North and other Australians must have spoken there about the value of medics to dryland farming because two major institutions, CIMYTT and FAO, supported projects in the 1970s to see what would happen when the medic was introduced into the existing farming systems in Syria, Tunisia and Algeria. In the intervening period the price for fresh sheep meat in the region had begun to rise substantially due to the higher incomes generated by oil wealth. The figures quoted in Chapter 1 show that eventually in 1989 sheep meat prices in Algeria went on to become probably the highest in the world and the dependence on imported grain for livestock as well as imported meat had grown exponentially since the Trumble and Oram handbooks had been published. Cereal yields had not improved to any extent.

Dr E.D. Carter (an agronomist at the Waite Institute in South Australia) prepared reports for CIMYTT in 1974, FAO in 1975 and ICARDA in 1978 on the potential of a medic farming system to the countries in the region. His was an eminently practical approach. He explained the importance of shallow cultivation to the success of the rotation and the need for scarifiers to carry it out and emphasised that it was essential to use implements designed for the specific function. He stressed the need to train farmers in the use of scarifiers and in the type of grazing regime best suited to the profitability of the system, and also the need to educate technicians working in the extension services about the technical and scientific information that provided the rationale for the management of the system. His reports are lucid, extensively documented, and represent an investigative mind firmly based in the direction of the practical farming requirements of the system (Carter, 1974, 1975, 1981).

CIMMYT supported a project in Tunisia to introduce medic pastures in place of fallow, FAO supported projects in Algeria and Syria for the same purpose and ICARDA explored and monitored the possibilities for fallow replacement that legumes (both crop and pasture) may provide. The Australian Governments established their projects in Libya and began negotiating projects in Jordan, and later in Iraq and Tunisia. During this period the term 'ley farming' was adopted by Australian technicians as a description of the rotation of cereals and medic pasture and it can be argued that it was a misleading title, confusing the system as it did with the rotation of cereal, legume or root crop and sown pasture used in Europe. Foreign technical experts continued to visit Australia to discuss with colleagues the medic/cereal system and to visit farms to see parts of it in operation.

The 1979 FAO conference

The result of all this activity and its influence in reinforcing knowledge about the medic farming system in the Northern Hemisphere is apparent in the proceedings of a seminar supported by FAO and held in Amman in May 1979 (FAO, 1980). The participants were working on programs related to 'rainfed agriculture', mainly in North Africa and the Near East. One can also gauge from these proceedings the progress that had been made towards replacing fallow with either crop or pasture.

(a) Research results

Reports of research undertaken reiterated what most technical and scientific experts already knew.

- Bare fallow was wasteful in terms of lost production, and costly because it needed a large investment in machinery and implements and it took a great deal of time, fuel and effort. It had no effect on moisture conservation in dry areas, and was destructive to the soil, and it was not contributing to increased yields of cereals.
- There was a growing body of work to show that cultivation to depths as little as 6 cm was cheap, effective, and did not lower yields of cereals. Comparisons with deep ploughing showed that the benefits claimed for deep ploughing *vis-à-vis* shallow did not exist.
- Superphosphate was proving a useful fertiliser for cereals and the placement of cereals and fertiliser together in the seedbed contributed to

more efficient germination of seed and uptake by the plant of fertiliser.
- Weed control was a problem in crops and deep ploughing and bare fallow did not deal with it efficiently.

Still the existence of a means of carrying out cheap and effective shallow cultivation on cereal farms in Australia did not rate a mention in spite of Carter's reports.

And, in spite of good reports of the growth of medics on the early projects and its exploitation in a profitable manner on farms in Libya, the search for a replacement for fallow on most research centres remained concentrated on grain legumes – lentils and chick peas – sown in fields using deep ploughs and cereals fertilised with nitrogen fertiliser. Yet on the farms the reception to these legume crops by farmers continued to be unenthusiastic for much the same reasons that Oram had listed in 1956.

- Grain legumes proved often beyond the means of farmers because of the need for more machinery and time with which to carry out the sowing program.
- Lentils and chick peas, never widely taken up, were declining in use because of the lack of machinery with which to harvest them.
- Vicia (vetch) hay was the next best option but the cost of specialised haymaking equipment was proving a barrier to widespread adoption.

The few research centres that had run trials with medic pasture on the other hand reported that

- Medic pastures had shown that large gains could be made in availability of feed for livestock. The so-called 'Australian varieties' of medic had been sown and had grown well in most countries in the region.

(b) Reports of national programs

Outside the research centres there had been some broad acre sowings of medics. Algeria had begun a medic/cereal program in 1972/3 and had established 30 000 ha of medic, Tunisia had established 17 000 ha since 1971 and planned a further 5000 ha each year until 1981, Libya had established 27 000 ha in the Tripoli region and planned a further 10 000 ha each year and, although the conference did not receive a report about it, the Seedco farmers, the demonstration farm in the Jebel el Akhdar and Benghazi Plains project had also established 80 000 ha on farms and more on the rangeland.

- From Iraq to Morocco the trials with medic had resulted in good first year pastures.

- The average rainfall in the zones where it had done well ranged from 150 to 350 mm.
- In Libya, the Gefara Plain project established in the 200 mm annual rainfall zone south of Tripoli, achieved yields of two tonnes of wheat after medic and a stocking rate of two/four sheep per hectare for the entire year grazing medic pastures.
- The contribution to soil nitrogen after medic pasture was measured, and the improvement in soil fertility resulted in markedly increased yields and productivity. Sheep everywhere had improved their nutrition substantially when grazed on medic pastures and supplementary feeding had been rare.
- The cheapness of the system *vis-à-vis* the other alternatives proposed was being quantified.
- Local varieties of medic *in situ* had been identified and were being followed up with local seed production centres in mind. Seed production units to multiply varieties already growing on the centres were also set up or planned.

The results led the Saudi Government, combining with FAO, to organise a one year training program on Dryland Farming Systems in South Australia so that countries in the region could send their young scientists to become familiar with the underlying principles and operation of the medic/cereal system as it was used in Australia, to learn how to produce seed for trials and large scale demonstration, and to exchange technical information. Twentyeight graduates had already successfully completed the course and eleven more were to begin (FAO, 1980).

(c) Results on the farms – institutional pessimism

In spite of the favourable results being obtained, the emerging picture at conferences of the rate of adoption of the medic farming system on the farms was gloomy. But could this gloom really be substantiated? During the decade the Seedco farmers had worked with some hundreds of farmers in the El Marj district and they had together established continuous medic/cereal rotations on their farms. The grazing phase was not being paid sufficient attention, mainly because of the concentration on the tillage and seeding operation that required the Seedco farmers to train their counterparts in machinery use and maintenance, but cereal yields had improved as had soil fertility, the medic pasture was being grazed not cropped, and farmers were not buying in supplementary feed for their farm flocks.

The Algerian project had only had a short term (three years) and this was not long enough to establish a sufficiently comprehensive farmer education program to ensure adoption, but several farms had established a rotation and it continued past the life of the project. No extension of the first phase had been agreed and so the national institutes followed on with their own research into the system.

The Gefara Plains project in Libya had successfully established a continuous medic/cereal rotation on large stations that continued for a decade, but the allocating of individual farms from these sites had not yet taken place.

The Tunisian project had a short term (three years) and no extension of the first phase was agreed, so research programs reverted to the national institute but some of the Cooperative (UCP) farms which had taken up the medic continued during the decade and beyond to graze their sheep flocks on regenerating pasture and to persist with attempts to use discs and chisels for shallow cultivation.

The gloom tended to come from those technical experts who had worked on the Tunisian and Algerian projects and it stemmed from disappointment when many pastures failed to regenerate effectively after the first successful year.

Carter had provided a concise account of the requirement of shallow cultivation for the successful regeneration of the pasture, and the necessity for the properly designed implement to carry it out. However, only the Libyan projects followed this advice. On the other projects the farmers had to cultivate the soil with implements designed for deep ploughing. When 'modified' in an attempt to perform shallow cultivation, they did a poor job and were ineffective in weed control. The El Marj farmers and the Gefara Plains teams used scarifiers and they had no problems with regeneration. The simple connection between tillage malpractice and weak pasture was not always made. Many of the technical staff, concerned at the lack of on-farm adoption, began to look elsewhere for what they called 'failure of the system'.

They fell into two groups, those who blamed 'management of the system' and those who began to criticise the varieties of medic used, claiming that the Australian varieties were responsible for the failure of the system as they were not 'adapted' to the region.

(d) Management failures

Some of the reasons for management failure originated on the pilot farms where there was confusion about the way in which the system used by

Australian farmers was operated. An example is the project carried out in Syria in the 1970s by a team employed by FAO.

Details were given in Amman in 1979 of the establishment and management of a project established in 1977 by FAO in North East Syria using the Himo Research station at Kashmili to 'intensify and diversify the existing wheat/fallow/wheat rotation with various legumes now under trial, introducing sheep into the cropping system as an integral part and upgrading the poor selection of agricultural implements'. (FAO, 1980, p. 50). The site was in classic dryland farming country – a winter rainfall that varied from 200 to 600 mm, occasional frosts and snow in winter, summer temperatures of 40 °C, and a drought approximately every three years. The common rotation was wheat/bare fallow/wheat/bare fallow and production was highly mechanised. Most farm households had some livestock (sheep, goats and cattle) but the major flocks came onto the farms at the end of summer and went when the stubble was ploughed in. No grazing was available at other times for these outside flocks because of the bare fallow. Farmers had (on the advice of technical experts) tried growing lentils and chick peas that had to be hand harvested and there were some attempts to produce vetch hay in place of bare fallow but 'this activity has seriously declined since 1976'. Further diversification remained problematical. The station land consisted of 320 ha most of which was divided between cereals and grain legumes used for fodder bales, silage, and some grazing. Another section was used for cereal seed production and a separate section was set apart for a medic trial.

(e) Seed bed preparation and time of sowing

The seed bed for the medic was prepared with an initial disc ploughing at the end of summer followed by several passages of disc harrows. A program of deep ploughing was used to prepare the seed bed for the subsequent cereal crop. The medic seed was sown at rates varying from 6 kg to in excess of 12 kg/ha into cloddy seed beds.

The difficulties encountered by farmers when trying to stretch time, machinery and implements to prepare land both for a cereal crop and a grain legume crop seem also to have been experienced on the station and may explain the fact that much of the medic was sown two to three months later than the cereal crop, and later than the grain legumes that had priority. Much of the medic was not sown until mid- or late winter. It was well known by then that time of seeding has an effect on yield. One would have thought that this was just as important for a medic pasture as it was for a cereal or grain legume crop. The effect of early sowing on the productive capacity of the medic is important. Late seeding puts seed production and regeneration in jeopardy. When sown just before or immediately after the autumn rains the result is early germination and the pasture is able to benefit from early spring warmth to produce abundant seed. With good supplies of seed the medic is able to regenerate adequately after a cereal crop in subsequent years and the farmer need not cultivate except for his cereal crop. Because of the cultivation and sowing program followed on the project, few of these benefits of medic could be demonstrated.

It is hard to believe that the project budget could not include finance for a relatively cheap scarifier and a combine seeder. The scarifier and combine seeder could have been used not only for the medics and cereals but also for legume crops where rapid cultivation in the autumn is also critical to good yield. Such a program may well have enabled the station to interest the farmers in improved tillage programs for all three alternatives. The Libyan projects had already established the value of employing experienced medic farmers to initially sow the medic and the subsequent cereals on the Libyan farms using a scarifier and combine seeder and this proved to be an effective way of improving tillage there. The sowing techniques used on the Syrian station were inimical to success. That it survived is a tribute to its harmony with the environment, rather than to the management that it received.

(f) Sowing and managing the pasture

If the intention was to demonstrate medic pastures then the initial sowings did not show the way to do it.

An experienced medic farmer would have sown 10 kg/ha of medic seed joined with 100 kg/ha of superphosphate. He would have sown the medic on its own so that a strong pasture would grow in the first year thus guaranteeing an ample seed bank that could survive grazing and the intervening cereal phase and still regenerate strongly in the third and subsequent seasons.

Most medics on the Kashmili station were sown with wheat or vicia (vetch) – then rolled with a heavy bar dragged behind a toothed roller; 100 kg/ha of superphosphate was broadcast over most of the sowings, but some received various other fertilisers – 'even ammonium sulphate'.

All medics sown on the station were varieties imported from Australia and all grew well and produced good seed sets. The only 'failure' was when the variety Hannaford (Barrel) was mixed with superphosphate in February and sown at a rate of 6 kg/ha into a growing wheat crop that was far advanced. It was not poor germination that caused the failure – indeed, germination was good – but competition for moisture with the standing crop. Medics sown as late as March were 'little affected' but Vicia sown later than December was a 'waste of seed, labour and fuel'. The yield from six varieties of commonly used medic sown alone was superior to the same varieties sown in conjunction with 30–40 kg/ha of durum wheat seed. In addition to the mixed sowings on the project site 'seven species (of medic) are sown this season in multiplication plots, 2 ha each . . . in a separate field with emergency access to irrigation' (FAO, 1980, p. 54).

This use of medic as a mixture with forage crops such as vicia (vetch) and wheat prevented any assessment of the savings and productivity associated with a medic pasture. The demonstration on this station was not that medic pasture was a viable alternative on its own to fallow but that medic was a valuable component in a forage crop replacement for fallow. This was an alternative that farmers had already rebuffed.

(g) Exploitation of the medic

Grazing of the mixed medic and vicia (vetch)/cereal crop on the station proved difficult because it was impossible to get enough sheep to exploit it and so no fixed sheep numbers per hectare could be calculated. Top mowing was necessary to control weeds due to insufficient grazing. Of the sheep that did graze it was recorded that in a handpicked group of 90 lambs the average weight gain of 15.2 kg over 66 days, represented a gross margin of $12 300 due to 'high protein medic pastures and green Vicia stubble'. The team were appalled at the waste of feed because of the absence of sheep but hoped that in the subsequent season with better organisation the 'gross profit could be quadrupled without supplementary feeding'. In February of the first year, 500 sheep were put onto 50 ha of mixed vicia (vetch) and medic and the team hoped to increase the number to 2000 by the end of March or early April, graze them again in May–June before the wheat harvest and then tail the numbers off to 500 or 250 in August. The team concluded that 'Perhaps one of the biggest problems will be convincing farmers to purchase sufficient sheep to profit from the increased production' (FAO, 1980, p. 50).

The hope of trying to 'convince farmers to purchase sufficient sheep to profit from . . . increased production' in the manner followed on the station seems a curious one. Farmers with small to medium size farms tend to have a relatively stable number of animals in their farm flock and their preference is to have steady and reliable sources of feed for them. Most farmers find it too costly to buy in large numbers of animals for a short period to take advantage of lush pasture and then sell them quickly when feed falls off. Such a strategy sets up high risks as buyers are quick to take advantage of sudden releases of hungry animals onto the market. The grazing regime on the station tended to be the stop/start method usually followed for a forage crop known colloquially as 'the vacuum cleaner' method. Medic pasture, when managed properly, is grazed steadily by a carefully balanced number of sheep for most of the year and only in mid-winter and at flowering time in the spring is there need for caution.

(h) Haymaking

It was noted on the station that the inclusion of Snail medic with vicia (vetch) increased hay production by 'more than 40% . . . if plot weights can be relied on . . . [and] . . . although hay production is a current priority the grazing medics are being observed with interest' (FAO, 1980, p. 57). The continuing resistance by farmers to the adoption of haymaking because of the cost of specialised machinery suggested to the project team that future programs should 'in many cases eliminate a hay species in favour of a grazing species' (FAO, 1980, p. 52).

(i) Extension message

The project report begins with a description of the general conduct of the medic/cereal rotation used in Australia. The actual practice on the station site remained far from that description. The contrast of savings available to the farmer from shallow cultivation and a regenerating pasture *vis-à-vis* the costs of sowing, harvesting, conserving and feeding out hay and grain legumes was not evaluated and used to develop an extension message designed to encourage farmers to adopt medic either as forage or as pasture.

(j) Networks and experience

The Syrian project appears to have taken place in a vacuum. The planning of the project took no account of the findings of the FAO Algerian project that shallow cultivation was important to the successful operation of the system, nor did it seem to have been influenced by the experience in Libya of either Seedco teams or the Gefara Plains teams. The farm practices observed by foreign technical experts during their visits to South Australia seem to have had no effect on the project formulation or evaluation exercise.

The project in Syria used a pilot farm to demonstrate alternatives to fallow to farmers but developed grave faults when the range of alternatives proposed were not properly understood, and therefore not clearly demonstrated. Alternatives already proven unacceptable to existing farmers because of environmental or material conditions were not discarded and research and trials redirected towards those that appeared better suited.

Twenty years of research in the region (let alone fifty years in South Australia) to identify useful techniques against impractical ones seemed a waste of time if no account was taken of it. The contrast between the complex medic regime followed on the Kashmili station and the simplicity of the Seedco project in Libya was stark. In the 1980s expertise in shallow cultivation and grazing management would become even more important as interest in a medic/cereal rotation to replace fallow gathered strength and governments asked for extension-based projects to enable the system to be demonstrated and explained to their own farmers. The institutions were expected to provide the expertise for these projects. How well did they respond?

Conclusion

Between 1953 and 1980 the institutions and agencies explored many aspects associated with the replacement of bare fallow with medic pastures. They showed a lack of appreciation of the savings that this system offered dryland farmers both large and small. They simply absorbed medic into their own favoured system of a grain or forage crop in place of bare fallow. As the farming operations and grazing management needed for a pasture/cereal rotation continued to baffle them they declared the system too complex for the farmers in the region and turned their attention to the biological aspects of the medic plant.

10

Institutions, agencies and medic – 1980–93

Introduction

Until the various projects began operating in the early 1970s the agencies and institutions had been discussing an Australian system that had not been tried on farms in the region. In the decade to 1980 a great deal of experience was gained about the system on local farms and by 1980 various models of introduction and adoption were available on which to build. The problems of adoption were remarkably similar in each country – poor regeneration of the pastures after the cereal phase and difficulties in understanding the grazing management required to exploit the feed available. Using this experience and the resources available to them the agencies and institutions should have had a reasonable chance of leading the way to a vast improvement in dryland farming in the region. How did they cope?

New projects

The Libyan projects were convincing proof that a medic pasture could be managed by farmers in the region to provide a stable and productive alternative to a bare fallow and that expert farmers were an effective means of transferring the operational know-how of the system to other farmers. The Libyan projects also demonstrated how, if farmers were given appropriate farm implements, the tillage problems were rapidly overcome. After 1980 other governments in the region wanted the rotation adopted by their farmers – not simply demonstrated behind the fences of the project site or research farm. Their enthusiasm contrasted with the doubts and frustrations being expressed by technical experts. It was unfortunate that the experience of the expert medic farmers who had achieved success with their Libyan counterparts at El Marj was rapidly becoming marginalised. The

Tunisian Government decided to set up their own independent program centred on a livestock production station at Souaf using experience gained during the CIMYTT project of the early 1970s. Morocco sent a team to Australia to look at medic farms and they decided to undertake, with some guidance from the German agency GTZ, a large project called 'Operation Ley Farming'. Projects in Jordan and Iraq were negotiated with Australian governments who contracted them to Australian agencies. This new phase demanded experts who understood how and why shallow cultivation was used and also for experts who were familiar with a grazing regime that managed the resources of pasture and cereal stubbles profitably. Some of the projects did employ expert farmers but, in the main, they were used only to drive tractors on the stations and not to train other farmers or advise on project planning and operation. Departments of Agriculture and research institutions remained the major recruiting source for project staff and it was they who set the directions that the projects took. What they had learned from their experience and how they dealt with the medic system as a result is presented in publications and conference and agency reports.

Arguments and opinions

Dr Doolette contributed an essay to a Ford Foundation/Stanford University publication intent on improving dryland agriculture in the region. He wrote in 1980 that although the medic system was 'a simple one and easily understood by technicians and farmers' he had encountered problems in Tunisia with the 'conceptual problems that arise' because 'agencies, organisations and people become interested in [the system] without having had any significant experience' . . . and it is . . . 'hard to visualise the system and its management requirements without seeing it first hand' (Ford Foundation, 1980, p. 76). Doolette's claim that there was nowhere in the region where it could be seen in operation ignored the farms and development sites in Libya (which were visited in 1979 and 1980 by Iraqi, Algerian and Tunisian Ministers and their officials) and Domaine Chouhada in Algeria.

In 1982 ICARDA invited Dr J. McWilliams to evaluate its pasture research program and suggest a new one. The Carter report for ICARDA on the potential of medic pastures to the region had been available since 1978. The invitation to Dr McWilliams probably came as a result of Carter's confident predictions.

McWilliams trenchantly criticised the pasture programs being undertaken by ICARDA. They had, he said, 'no clear purpose and no sense of priority'.

They should be brought back to concentrate on the forage needs of existing farms in the region. His conclusions, gained from interviews with technical experts working on projects in the region, were that the major barriers to the system being adopted were inefficient weed control and a failure of regeneration after the cereal phase. Grazing the pasture efficiently had proved difficult. He recommended that ICARDA concentrate on selecting 'adapted' medic varieties, and assess the feasibility of a local seed industry. Tribes and villages should control their common grazing systems better and farmers should be able to sell the grazing rights to their pasture to other livestock owners (McWilliams, 1982, p. 14).

The list was like bird shot and it missed the essential problems which had been clearly enunciated in most reports. Someone had to make a scarifier available to the farmer and then demonstrate how the implement was used to cultivate the land for the cereal crop and how to graze sheep on the medic pasture and cereal stubble so that these could replace most of the currently purchased grain and straw.

Carter's earlier recommendations to ICARDA had concentrated on these practical aspects and he had stressed that it was also essential to educate technical staff in the practical as well as the theoretical aspects of the rotation. Without these measures the ubiquitous problems on the research stations and associated farms of burying the medic seed too deeply during the cereal phase and using the pasture as a forage crop would continue.

Australian civil servants and administrators, as well as technical and scientific experts involved in the Australian projects of the 1970s and early 1980s, were interviewed by Dr R. Springborg, who carried out a review of medic in the region in 1985. They were sharply divided in their opinion of the reasons why progress had been so slow.

The first group believed that medic should only be used on farms in the region as part of a forage crop. They believed that the medic/cereal rotation only succeeded in Australia because of the climatic aberration of relatively warm winters. The cold winters of the region spelt doom to any attempt to duplicate the system Australian farmers used. The system would always remain 'an idiosyncratic response to the highly favourable, temperate climate of Australia'.

A second group admitted that the cold winters of the region had not prevented medics producing good pasture, but believed no further development should take place before a course of careful experimentation was pursued into problems of pest control, specific varieties for specific sites, methods of weed control and so on 'lest by rushing ahead too rapidly disastrous failures discredit the Australian model irrevocably'.

The third group said that medic had produced lush pasture throughout the region. They said that the search for a 'perfect' operation of the system was unrealistic. No farmer (even in Australia) ran a technically perfect medic farming system and it was ridiculous to expect farmers in the region to achieve perfection before going ahead with an expanded program. It was the opinion of this group that the failure to continue the initial success and incorporate medic in place of fallow throughout the entire region was due to poor on-farm grazing and lack of scarifiers for the cereal phase. This group expressed despair that scientists/technicians would ever feel confident enough to move from 'the refuge of the experimental plot' to face the practical problems of management (Springborg, 1985, p. 10).

Although the Libyan farmers continued to operate successful rotations and pastures right up to 1983 when the Seedco team finally pulled out this was not used to support the arguments put forward by any of these groups. The Libyan experience had shown that when the farmers had the proper implements and good advice they could operate the system in the region. Its success should have made it the focus of institutions conducting research into replacements for bare fallow. In fact it was ignored, as was much of the work being done by the national institutes within the region.

In 1982 McWilliams noted that no medic research was being carried out by national institutions 'except where there is a joint project with expatriot (*sic*) scientists sponsored by GTZ, FAO, World Bank or Australian overseas projects' (McWilliams, 1982, p. 14).

At an expert consultation of administrators, technical and scientific staff of national institutions within the region, sponsored by FAO and held in Sidi Thebet in Tunisia in 1987, each delegate gave details of extensive medic programs and associated research in their own countries designed to replace fallow with medic pastures and increase production. The national institutions were carrying out their own trials to assess the potential of medic pastures and to explore aspects of its management. The identification of local varieties of medic was well underway. These delegates (who were working face to face with farmers in their own countries) complained that their work was hampered by

- a lack of extension material explaining the system and its operation,
- the absence of appropriate machinery for shallow cultivation,
- the absence of a local seed industry capable of producing commercial quantities of suitable seed,
- the absence of funds to provide these resources.

They were keen to see research into local varieties and germplasm continue but their priorities were

- a review of existing farm implements used for cultivation on farms and machinery centres and more research into the differences between deep and shallow cultivation,
- the provision of appropriate machinery for shallow cultivation,
- national workshops where technicians could be trained in the practical management of the system,
- some extension material,
- a study and analysis of the projects of the 1970s and in particular the Libyan farms and the degree to which they had influenced other farmers,
- a local seed industry to help cut the cost of commercially available seed,
- some means of communication between groups working on medic programs in the region,
- the establishment of a working group on medic pasture development in farming zones and rangeland to act as a core group to monitor progress (FAO, 1987).

The national institutions lacked funds to concentrate on these practical management difficulties they had identified and they required agency funds to help them deal with them. During the next decade this was the result.

Research undertaken

Although a medic/cereal rotation had now been successfully operating on individual farms and large development sites in Libya for a decade and medic pastures had produced ample and nutritious grazing during the establishment phase in Algeria and Tunisia to many technical experts the system still remained a black box.

An Australian scientist, Dr A. Clarke, suggested in 1980 that 'the use of legume crops to restore soil fertility may be limited because they compete with major crops for soil water' in the first place, and in the second 'the merits of crop/livestock versus cropping-only systems depend on biological, economic and managerial factors. Research on the first has been inadequate, and the second and third defy quantification' (S.A. Government, 1980, **1**, pp. 22–3).

Research and development

Many research programs were undertaken from the 1980s onwards. The question is whether they assisted the adoption of a medic system on farms or whether they retarded it.

(a) Bare fallow

Within the technical community both in Australia and in Rome, there were still some who remained loyal to the fallow/wheat rotation because, according to Dr De Brichambaut of FAO, bare fallow had a 'beneficial effect on moisture conservation where rainfall was below 300 mm' (S.A. Government, 1980, 1, pp. 4–6). Dr A.L. Clarke, of Australia, advocated bare fallow because it was 'an important component' due to 'increased water storage, including mineralisation of nutrients, weed control and convenience to the farmer' (S.A. Government, 1980, 1, pp. 22–3).

Results of trials and experience presented by many others continued to show that bare fallow in semi-arid conditions was not wise. There were better and more environmentally sound ways of achieving weed control and plant nitrogen. Even Dr Clarke admitted that the practice and duration of bare fallow in Australian cereal areas had diminished because of its adverse effect on soil fertility (S.A. Government, 1980, 1, pp. 22–3).

In 1989, FAO's pasture section was still championing bare fallow for the region because of 'moisture conservation, weed control and organic matter mineralisation that release nitrogen for plant use' (ICARDA, 1993, p. 16).

(b) Shallow cultivation

As part of the reasons for rejecting bare fallow, there was a general consensus that the high cost of deep ploughing and its resultant poor seed bed were a disadvantage to the dryland farmer. An alternative type of tillage was needed. At the same time, experience was showing that medic pastures were not regenerating after the cereal phase because of the conventional tillage being carried out on farms. Could the adoption of shallow cultivation solve both problems? There were some results that seemed to indicate that this may be the case. On Libyan farms scarifiers were being used and farmers were benefiting from the efficiency and low cost of the operation. The Jordan Cooperative Organisation carried out its contract services with scarifiers because it was cheaper and quicker than ploughing. On the project sites in Iraq the Australian teams used scarifiers demonstrating the benefits there. On most research centres, however, comparisons of depth of ploughing continued to be made with 'adapted' chisels and disc ploughs instead of scarifiers and although most showed that yield was not lowered by shallow cultivation, the benefits of good weed control and fewer passages over the ground to break down clods were not apparent.

It was not only the farmers who wanted to practise a medic/cereal rotation who were in need of a scarifier. Farmers who wanted to fallow, or crop cereals continually, or grow grain legumes or other fodder crops, were also suffering the liabilities of long cultivation programs, cloddy and uneven seed beds and poor weed control.

In spite of finding that deep ploughing on the Tunisian farms had buried the medic seed too deep for regeneration, Dr Doolette, for example, did not think farmers needed to purchase any special equipment for their tillage program (Ford Foundation, 1980, p. 76).

Yet another colleague reported that trials in Algeria had tested a scarifier against ploughs and discs and the scarifier had speeded up seed bed preparation 'as much as fifteen times'. At El Khroub research centre a trial had contrasted the conventional sowing program using plough and discs in separate seeding and fertilising operations with the combine seeder and scarifier that did all the operations at once. The combine seeder and scarifier trial produced better yields and better weed control in both a good year and a bad year. The writer pointed out that 'More efficient seedbed preparation equipment for both dry and wet soils will do more for timely seeding than any other factor' (Ford Foundation, 1980, p. 49).

Dr Pattison in Algeria found that the commonly available machinery for tillage during the cereal phase did not enable farmers to successfully control weeds and as this was the only means of weed control available to them the second year pasture was poor and its seed set low (S.A. Government, 1980, 2, pp. 111–14). The project team were forced to use discs and ploughs because these had been supplied to the project and no others were available.

In a paper presented to the Adelaide conference, two Algerian technicians, Adem and Benzaghou, reported results obtained when a scarifier and combine seeder had become available:

- weed control had been good using shallow cultivation carried out with a scarifier,
- cereal yields were better from seed beds prepared by scarifiers because seed beds were of better quality and timeliness of seeding could be optimised,
- shallow cultivation meant a reduction in the capital investment needed for machinery because
- scarifiers only needed wheeled tractors to pull them instead of the heavy crawler tractors required for the deep plough,
- and the cost of seed bed preparation was lower because it could be carried out more rapidly (S.A. Government, 1980, 2, pp. 170–4).

A machinery salesman at the same conference pointed out two rarely mentioned benefits of the scarifier to technical experts attending the same conference:

- its use enabled the medic seed to regenerate without difficulty and
- the scarifier was robust and simple and did not often need spare parts – a great benefit to cash-poor farmers isolated from machinery centres (S.A. Government, 1980, **1**, pp. 26–8).

In his recommendations to ICARDA in 1982 for future research programs pertinent to pasture, Dr McWilliams (although acknowledging the 'inappropriate' tillage that affected pasture regeneration) did not suggest a program to contrast the results obtained after using a scarifier and a disc or deep plough (McWilliams, 1982, p. 14).

In 1984 ICARDA established a research program to evaluate attempts to operate a medic pasture/cereal rotation on a group of small farms in the vicinity of the station. Problems with regeneration of the pasture after the cereal phase (in which cultivation was carried out with a locally used 'ducksfoot cultivator') were sheeted home to rough seed beds, overgrazing, inadequate grazing, or too many weeds. The response was to advise farmers to sow 30 kg medic seed per hectare rather than 10 kg in the hope that the initial dense growth of plants would overcome the difficulty of the rough seed bed and capped soil as well as reducing weed growth and providing so much pasture that overgrazing would not seriously diminish the available seed bank.

In 1985 Dr Cocks of ICARDA proposed that 'it may be best to simply introduce a concept in this case rotation of cereals with pasture then by limiting resources to those already available encourage farmers to build their own systems around the concept'. He believed that farmers could adapt the local tillage equipment thus ensuring that the medic would not be ploughed in too deeply (Cocks, 1985).

In 1989 at a workshop in Perugia sponsored by ICARDA, Dr Jaritz and Dr Amine of Morocco reported that by the end of 'Operation Ley Farming' there had been a decline in the area of medic, largely due to poor regeneration. Their description of the operation of the rotation revealed a high incidence of deep ploughing, tillage being carried out by disc ploughs 'adapted' to perform what proved to be an inefficient form of shallow cultivation, and attempts to control subsequent weeds with herbicide applied at a time injurious to the pasture. They reported that in every instance on the Had Soualem Station regeneration of the medic had suffered after the cropping phase, but that, in spite of this, the medic pasture was 'highly valued because it allowed a substantial reduction in concentrate feeding'. They concluded, not that the farmers should be able to buy scarifiers before proceeding further, but that

structural aspects such as farm size, status of land ownership; presence of fallow land, the existence of competing crops and the system of animal husbandry are as important as technical aspects such as proper stocking rate and shallow cultivation in the cropping phase of the rotation (ICARDA, 1993, pp. 35–41).

At the same conference, FAO delegates were pessimistic about the value of medic pastures in the region for a number of reasons; among them was their opinion that the system was too difficult for most farmers because it demanded 'complex sowing techniques that require sophisticated machinery available to few farmers' (ICARDA, 1993, pp. 15–22).

In 1993 the ICARDA 'Network' newsletter reported that as part of the WANA 'benchmark' program they were supporting in Algeria to encourage the use of a medic/cereal rotation it had been discovered that farmers believed that deep ploughing was necessary and in response a trial was being undertaken to 'see if deep tillage was necessary' (ICARDA Newsletter, 1993).

The relevance of shallow cultivation to the solution of farmers' general problems remained unremarked and unexplored by the institutions. It seemed that by the end of the 1980s the use of a scarifier had become so identified with a medic farming system in the minds of technical experts employed by international institutions that they could not conceptualise its use except as a not very important part of a rather mysterious medic/cereal rotation. They simply ignored the way the Libyan farmers, when using scarifiers, achieved rapid, cheap cultivation, got good weed control and succeeded in regenerating their medic pasture. They also ignored the evidence of the projects in Iraq where, behind the fences at Erbil and at Telafa, the project teams had used their imported scarifiers to achieve good weed control, and level seed beds for medic and for cereals, and regeneration of the pasture was good and trouble free.

In general, the technical experts saw no need to ask that scarifiers be supplied, but preferred to suggest manipulations of the medic plant, rather than the lack of scarifier, as the solution to this major problem of deep ploughing that bedevilled the farmers. Why were they so blind to the efficiency of the scarifier? A review of the experiments carried out to compare shallow cultivation with deep ploughing suggests that most researchers were not aware of the difference between a chisel and a scarifier and thought that any tyned implement was suitable for shallow cultivation.

The scarifier is not common outside of Australia although it is now manufactured by international companies. An understanding of the reasons for the different design of each implement is essential to carrying out any useful comparison of its operation. Adapting a plough or disc designed to efficiently turn soil over and penetrate the soil deeply is not an adequate substitution for a scarifier that is specifically designed for shallow cultivation. Attempts by farmers to maintain medic pastures in Morocco, Algeria and Tunisia using disc ploughs 'adapted' to the purpose of shallow cultivation during this period resulted in the eventual depletion of the seed bank (usually after five years) and poor weed control that contributed to low cereal yields. Had the scarifier attained the distinction of being purchased by national machinery centres in the same way that they routinely purchased disc ploughs and chisels, the problems associated with deep ploughing may well have been solved.

(c) Grazing medic pastures

If the regeneration of the pasture after the cereal phase was a ubiquitous problem, the grazing of the medic pasture was not far behind it. The project reports reveal that much of the problem arose when medic was sown in conjunction with a forage crop and then grazed like one. Those unfamiliar with medic pastures simply took the seed and incorporated it into the rotation as part of the forage crop that was familiar to them and it was grazed in the usual manner. The Australian teams in Libya (who did know how medic pastures should be used) had been forced to turn large quantities of the medic into hay when lack of sheep on the project sites prevented adequate grazing taking place. In Jordan, the early attempts by the Australian team to manage the grazing of the pasture were abandoned in favour of a forage of mixed medic and vetch and hay production. In Iraq, medic pasture was the basis of a successful community project where it was grazed by both village and nomadic sheep but most medic on the project sites was made into hay. The South Australians at Erbil in Iraq were forced to do this because they had trouble getting a suitable flock of local sheep. Elsewhere, nomads, lack of fences, and muddles about management were cited as reasons for failures to establish and graze medic pastures in an efficient manner.

In 1980, Dr Doolette reported trouble with his pastures in Tunisia because the farmers let their sheep overgraze the pasture and because they cut hay at the wrong time (Ford Foundation, 1980, p. 76).

Dr Pattison believed that the failure of the system to attain 'anywhere near its potential' in Algeria was because of 'lack of correct grazing management strategies and failure to integrate livestock in the cropping rotations'. Farm workers and 'sometimes the extension workers' did not understand the 'intimate integration of livestock with the cropping enterprise ... [there was] ... poor liaison between the separate sections of the administration who are running the cropping and livestock enterprises ... '. The problem in Algeria had not been overgrazing but rather a 'scarcity of livestock' on many farms. It had been difficult to get shepherds to take the flock onto the pasture if it was a long way away from the bergerie. Fences would have to be used to ensure twentyfour hour grazing in order to benefit from the pastures. Dr Pattison had managed to achieve good weight gains (227.7 g per day per lamb) with a daily regime of eight hours on medic pasture. He warned that bloat could occur unless sheep were fed a little dry hay before being put onto lush pasture after a night of fasting. He also mentioned the need to find an accommodation with nomad flocks that came to the cereal zone at the end of the summer to graze cereal residues (S.A. Government, 1980, 2, pp. 111–14).

These fears about nomadic incursions damaging medic pastures were later noted by Dr McWilliams in his 1982 report to ICARDA, and he suggested control of grazing at tribal level and some means of providing farmers with the right to sell grazing rights for their pastures (McWilliams, 1982, p. 14). Both these measures already existed in most countries in the region.

Algerian delegates to the 1980 Adelaide conference reported a weight gain of 220 g of live weight per head per day for lambs grazing medic pastures. Medic's high crude protein content (19.80–26.10% of dry matter) and an energy value of 2515 UF (1 Unit Fourragière equals energy given by 1 kg of barley grain) compared to weedy fallow of 250 UF meant that the health of sheep on medic was much improved, and the cost of establishing each hectare of medic pasture (243 AD) provided adequate feed for five ewes and an estimated margin of 'about 40 to 50% on the investment'. They also suggested that fences may be needed to control the large numbers of sheep that would eventuate once the vast areas of medic pastures they planned to establish were providing the amount of feed expected (S.A. Government, 1980, 2, pp. 118–25).

Australian technical experts who had worked on the demonstration farm in Libya reported that 'farmers who had sown medic pastures recognised that these pastures provided excellent grazing and conserved fodder for their livestock'. They considered the major difficulty in adoption lay in the absence of an official extension service to take the message of medic pastures to other Libyan farmers. They thought that fences would be a help to grazing management (S.A. Government, 1980, **2**, pp. 147–50). El Marj farms had boundary fences and young shepherds looked after the flocks grazing the medic pasture growing side by side with the cereal crop.

The Western Australian Libyan project during 1974/83 had selected a nucleus flock of local breeds that grazed medic pastures and cereal stubble to good effect in the cereal zone and on their rangeland site at Ajulyat. They published the results (Allen, 1979, pp. 5–9), and it seemed that this was a path worth pursuing in the region, but no further work was done to explore the possibilities.

The Springborg review in 1985 of the Australian projects that had then concluded reported a belief that poor grazing management had caused many to lose faith in the relevance of the system to the region (Springborg, 1985).

In 1985 the grazing trials on medic pasture at Aleppo in Syria had given excellent results. Sheep had grown well, increased their milk production, body mass was maintained even in summer and savings had been made of previously purchased fodder. Cocks tested some of Carter's assumptions made in 1978 that 3 sheep/ha could be expected to be maintained all year round in country similar to that surrounding Aleppo. The ICARDA trial grazed 5 sheep/ha on first year medic and Cocks claimed that 'regenerating pasture is sure to carry more' (Cocks, 1985, p. 14).

Sixteen major institutions were represented at an International Conference on Animal Production in Arid Zones (ACAPAZ) held at Damascus in 1985 and sponsored partly by two Arab institutions, the Arab Centre for the Studies of Arid Zones and Dry Lands (ACSAD), the Arab Organization for Agricultural Development (AOAD), and GTZ. The majority of papers (whose authors represented the major international institutions involved in development in the region) were concerned with breeding, animal health and studies of livestock feeding preferences. One paper reported the results obtained on rangeland with medic pastures and proposed further work to extend the use of medic as a means of rehabilitating rangeland and making it productive (ACAPAZ Proceedings, 1987, pp. 763–80). It was resolved to carry out a study of the Australian/Libyan projects that had introduced medics to Libyan dryland farms and rangeland. This resulted in 1988 in the El-Akhrass and Wardeh report quoted in Chapter 5 (ACSAD, 1988).

The Moroccan 'Operation ley farming' project in 1986 found that farmers tended to use the medic as a forage crop not as a pasture and that this caused overgrazing during the pasture phase. The medic was valued by farmers 'because it allowed a substantial reduction in concentrate feeding'

(ICARDA, 1993, p. 35). There had been no problems with 'common grazing' or incursions by nomadic flocks. On Had Soualem Station there had been an instance of ewe infertility that the farm manager had suspected may have been due to grazing first-year medics on the station. The fecundity rate rose again in subsequent years and the regenerated pasture gave no further sign that medic was lowering ewe fertility (Jaritz & Amine, 1989, p. 16). This suspicion of medic-induced infertility (in newly sown pasture) is unique to the region. No other suspected cases have been reported. Research comparing Libyan and 'Australian' lines revealed no danger (ACSAD, 1988, p. 33). The experience in Australia has been with clover-induced (not medic) infertility among ewes and an account of this is contained in a study of a farm development scheme on Kangaroo Island (Nunn, 1981).

At the 1987 expert consultation in Tunisia, delegates from Jordan, Tunisia, Syria, Algeria and Morocco claimed that grazed medic pastures grown under technical supervision had provided better feed and returns over an entire year than any other form of crop or forage. The problem they found was the lack of experienced personnel to demonstrate to farmers how to manage and graze the pasture to best effect. Both the Tunisian and Moroccan delegates reported cases of bloat when hungry sheep were put out onto green medic, but also reported that the feeding of a little hay prior to grazing had prevented further problems (FAO, 1987).

In 1989 the Australian team in Jordan were still ruminating on whether farmers should be advised to grow a vetch forage crop or simply a medic pasture. Vetch was preferred by the team because it

is similar to pulse crops, is native to the area, and does not need special care during the season. However, in the long term, medic could be a good choice especially for farmers who own sheep and can practise direct grazing. However, medic requires special care in establishment, regeneration and animal/crop management. That some elite farmers have been successful in handling the medic/cereal system promises a good future for this system in Jordan (ICARDA, 1993, pp. 71–2).

After being in Jordan for nearly a decade, grazed pastures were still something 'to be attempted' even though they 'promised a good future'. The establishment of medic pasture and the grazing of it had only taken place behind a fence on land given by the university to the project team. No ordinary farmers were encouraged to emulate this.

FAO's pasture section in 1989 presented its conclusions about the problems with grazed medic pasture. Medic pasture was:

- the instrument of increased rat populations in Libya,
- the means of causing deaths of grazing sheep (within 15 minutes), through bloat,
- too difficult for farmers because pastures demanded 'sensitive techniques of sound pasture establishment and management',
- and it would not survive the cultural conflict with current Arab farming and grazing traditions (ICARDA, 1993, p. 16).

The identified problems with grazing once the institutions and agencies tried to apply their knowledge to individual farms had been either too few sheep or a difficulty in understanding the type of grazing required by a medic pasture.

A genuine pasture phase in the farming sense was hard to find on research stations and centres by the 1990s, yet no project or institution was without a continuous program to identify and select 'adapted' varieties of medics and test them with various types of inoculant. No international institution was supporting the use of medic pasture for rangeland regeneration and grazing. The Libyan rangeland successes on the Benghazi Plain and the Ajulyat site were lost in the past.

It is notable that many of the discussions of the grazing potential and grazing management of medic pastures within a medic/cereal rotation do not take account of the reliance on cereal stubble that is part of the system in Australia. Most concentrate on the productivity of the medic pasture alone and even here it is often only the production of the green pasture rather than a combination of both green pasture and dry straw and pods. The idea of the rotation is to grow medic and cereals on a part of the farm in each year rotating the pasture and cereals in roughly alternate years (depending on the farmer's judgement of his particular requirement). Thus the sheep have access to both pasture and stubble and it is the manipulation of the medic both green and dry and the cereal stubble and straw that enables the farmer to supply his flock with year-round nutrition from his own farm.

(d) The search for 'adapted' varieties of medic

A failure of commercially available medic varieties to grow did not appear on the farmers' list of reasons as to why they found it difficult to adopt and continue a profitable medic system, and yet during the 1980s the major reason given by technicians and scientists for the supposed failure of the system was a lack of adapted varieties.

In 1976 Webber reported that local varieties had re-established themselves and dominated the original Australian-produced cultivar sown on Algerian cereal farms. He suggested that no matter what variety of medic was sown to establish pasture, the local varieties would come to dominate once deep ploughing ceased and a farming system encouraging to the extensive population of spontaneous medic was in place. During the same period an opinion was expressed that the Australian-produced cultivars may not grow successfully above 600 m on the high plateau. Subsequently this was proved not to be so.

In 1978 Carter wrote that enough tested varieties of medic existed to establish trouble-free pastures throughout the major part of the cereal zones of the region and much of the rangeland. He was more concerned about the absence of information about shallow cultivation and grazing management.

In Libya there were no problems. Medic even regenerated well and was grazed in successive years on rangeland in a lower rainfall zone than was considered its habitat in Australia at the time. In this case, it proved to have a greater range of adaptability in Libya than it had in Australia. The Western Australian team took medic seed from this 100–150 mm zone of the Libyan rangeland to Australia and later produced the cultivar 'Swani' that has provided a means to extend medic pastures further out in their own marginal zone. During the mid-1970s Gintzburger and Blesing identified 1000 varieties of medic in various zones of Libya.

There were no problems with germination or growth of the original varieties of medics used in Tunisia, yet reports from the first project provide one of the first examples of the way in which the search for 'adapted varieties' came to distract attention from shallow cultivation and grazing management. Dr Doolette, who had directed the CIMYTT project in Tunisia, wrote in 1980 that difficulties with management were the major problems encountered in Tunisia, but at the Adelaide conference in the same year he changed his emphasis, claiming that a suitable system for the region depended upon research to find 'adapted varieties', ensuring that they were inoculated with appropriate rhizobium, and that they produced a good supply of hard seed (S.A. Government, 1980, **1**, pp. 38–9).

'Operation ley farming' in Morocco in the mid-1980s reported no problems with either germination or growth of the medics used. This was surprising as the inclusion of cultivar Robinson Snail (a medic used in the higher rainfall zone in Australia) for a project in Morocco that included medium to low rainfall bands could well have caused failure, but it did not. The eventual failure of regeneration that led to discouragement related to cultivation techniques and grazing management.

In Jordan in the 1980s there were no reported difficulties with germination or growth using the commercial cultivars distributed to farmers. Most farmers used the seed as an addition to a forage mix. Reports of productivity increases because of medic were universal. The problems with growth on trial plots at research centres was much commented upon by the technical staff but the project reports indicate that this was due more to late seeding, invasions of unauthorised ploughing and general muddle than scientific certainty.

The two projects in Northern Iraq in the 1980s demonstrated that medics grew even in conditions of extreme cold. Growth slowed in the winter months but the pasture was abundant in early spring. It produced so much feed above grazing requirements that hay could be made and stored for the lean period. This was a management model for cold areas to add to the one recorded by Pattison in Algeria. The new cultivar, Circle Valley, produced commercially in Australia and used in 1982/3 on the South Australian project at Erbil, responded well to the climate there. The Western Australian team in Iraq also found Circle Valley and Cyprus useful in responding to the cold problem. The major difficulty in establishment of the available varieties in Iraq was that in places the soil was so exhausted that the nodulation of the medics was poor and rhizobium had to be inoculated in order to kick start the process again. When this was done the pasture production was good. The FAO project in Syria experienced no problems with the 'Australian' varieties used during the 1970s.

Later, ICARDA, following the recommendation from the 1982 McWilliams report that they establish a selection program to identify 'adapted' ecotypes for high altitudes, sent a group to explore and they found 1000 varieties of medic in various climatic zones in Syria in 1984/5. The group of farmers in Tah village, discovered and subsequently monitored by ICARDA, were on their own initiative grazing regenerating medic pastures (*M. polymorpha*) spontaneously growing on parts of their farmland. Trials were undertaken on the ICARDA station of cultivars and varieties, some locally collected and some available in commercial quantities from Australia. The Australian varieties (with the exception of Circle Valley) were badly affected by frost in the first year of the trial and as a result ICARDA decided to concentrate on multiplying and distributing two Syrian ecotypes *M. rigidula* and *M. rotata*, which seemed to withstand the frost better. The Tah farmers who were grazing their spontaneously appearing medics (*M. polymorpha*) were given seed of the new varieties. Samples were sent throughout the region to research centres for trials and they proved to be frost tender in some (Algeria) and less than spectacular in others. Halse reported in 1989 that *M. rotata* and *M. rigidula* appeared well adapted to the cold winter in Iraq, and *M. rotata* seemed not to be attractive to birds which ate some of the pods of other ecotypes (ICARDA, 1989, p. 9). In Tunisia and Algeria, the Australian produced cultivar Jemalong (*M. truncatula*) sown on farms survived frost and snow and produced abundant foliage and pods later in the season. Cocks wrote in 1989 that Jemalong and Harbinger survived frost in 'four years out of five' in Syria (ICARDA, 1993, p. 46).

The problem of altitude receded as more concentration was put on selection for frost damage and drought. Both these seasonal hazards are part of farming and the variability of their occurrence is immense throughout the region, and although farmers are grateful for discoveries that lessen risk the most one could expect of this research is that it might help enlarge the original mixture of seed available to the farmer for his establishment sowing of pasture. It is an exaggeration to say that without these varieties of seed the farming system cannot be a success.

ICARDA joined its selection program to a rhizobium production and inoculation program and advised all those sowing medics to inoculate them first. In 1989 at the Perugia conference, the ICARDA delegate claimed success in establishing a medic/cereal rotation on several groups of small farms and gave the credit for it to the ICARDA program of

• selecting new medics that resolve the problem of 'adapted' cultivars,
• using indigenous medics that avoid the rhizobia problem while at the same time providing rhizobium where necessary (ICARDA, 1993).

None of the delegates to the Sidi Thebet consultation in Tunisia in 1987 referred to difficulties with the survival of Australian-produced cultivars. Some did have difficulty in getting or paying for imported seed and many wanted a local seed industry to produce their own supplies.

By the early 1990s it had become an article of faith within the institutions that 'adapted' varieties are the key to success with the system. There is nowhere in the region where it has been demonstrated that growing an 'adapted variety' solves the problem of poor tillage and bad grazing management. As a risk reducing mechanism it falls short of hay made from regenerated medic in good years and straw from cereal crops stored for use when drought or frost lower the yield of expected crops.

Why was there such an obsession with 'adapted' varieties? In Australia medic is considered an exotic and technicians and scientists involved in projects were more familiar with the programs to select medics for specific conditions than with the broad application of medics on farmland. When they wanted to duplicate their Australian research objectives within the programs of the institutions of the region or make it part of a project, they fitted easily and their marginal relevance to the task of changing the farming system went unremarked.

The selection of specific cultivars and their bulking up by the seed industry in Australia did help to extend medic pastures out into more marginal areas – for example, Harbinger, that provided a medic for the semi-arid zone of the cereal belt. However, the greater part of the cereal zone was colonised with, or sown to, Barrel and Jemalong (*M. truncatula*) and improved cultivars of these varieties are still the mainstay of the system.

In the absence of a natural community of medics it is possible in Australia to sow a particular cultivar chosen for its environmental aptitude or higher productivity and have it maintain its dominance. In North Africa and the Near East the prolific natural population of medics means that an introduced cultivar, if deep ploughing does not intrude, will lose its dominance to the rapid colonisation of the spontaneous varieties that are naturally adapted. These spontaneously occurring medics will take over from the sown cultivar whether it is 'Australian', 'Syrian', 'Algerian' or whatever.

The early tendency to call Harbinger, Jemalong *et al.* 'Australian medics' caused national identity to become associated with medics. The 'Australian' medics were blamed for failures although the evidence does not support this. The programs to identify and select 'local' or 'adapted' medics pandered to this nationalistic feeling, and by the mid-1980s on research stations throughout the region one was shown medics that were 'Algerian' or 'Syrian' or 'Moroccan' as though they were intrinsically superior to the 'Australian' medics. This was a fertile field in which to sow the idea that an 'adapted variety' was needed before the medic system could be successful in the region. Of course, all medic came from the region. The Australian contribution had been finding out how to use it in a farming system and to develop a successful industrialised seed industry.

One of the problems with original trials of medic in plots in the region was that these were sown as 'pure' stands, whereas most farmers use a combination of cultivars to cope with seasonal variations in rainfall, summer heat and so on. The spontaneous medic consists of a mixture of varieties and ecotypes that wax and wane from year to year. The 'pure' stand is vulnerable to variations in climatic conditions, and sowing only one variety on a farm exposes the farmer to loss if frost comes at the wrong time. The commercial variety 'Barrel' that initiated the South Australian seed industry was in fact simply a collection of seed separated out from farmers' cereals and included several ecotypes and probably one or two sub-clovers. In 1989 ICARDA announced in Perugia that it was no longer advising that 'pure' stands of a particular medic cultivar be sown.

Once seed production units are successfully established in the region the use of 'local' medics as a source of bulk seed is sensible, but the route undertaken in order to arrive at this sensible decision has been long and tortuous and illustrates the institutional narrowness of view that can allow a relatively peripheral objective to obscure the more immediate one of establishing a better farming system.

(e) A seed industry

While the quest for an 'adapted' variety provided a smoke screen behind which to hide failures of management and the absence of scarifiers, the quest for a local seed industry to provide commercial quantities to farmers who wished to grow medic pastures was a practical objective and one that many desired. Yet, here again the institutional point of view was at variance with the facts.

In 1980, Dr Boyce of the South Australian Department of Agriculture told the Adelaide conference that a lack of a seed industry was the critical constraint to dryland farming improvement.

The history of the development of the Southern Australian dryland farming system has highlighted the need for the concurrent development of a well structured seed industry actively facilitating agricultural production. It is doubtful whether the highly developed farming system currently in use [in Southern Australia] would have been possible without ... an active seed industry (S.A. Government, 1980, **2**, pp. 139–42).

The evidence indicates that this is overstating the case somewhat. The farming system in its entirety was well established on many farms before the first commercial medic seed variety (Hannaford/Barrel) became available, but it is true that the medic farmers in the 1970s found that paying to import their own seed back from neighbouring states where the seed merchants were based was an invidious position and they set about establishing their own local seed industry.

The countries in the region were keen to have their own seed industry and they began to pursue this objective in the mid-1970s when Tunisia imported suction harvesters and undertook a seed production program.

A little later, the Algerians set up a seed production unit at Guelma and the Moroccans established a seed production unit in Western Morocco.

In 1982 McWilliams recommended that ICARDA carry out an investigation into the feasibility of a seed production industry (McWilliams, 1982, pp. 1–14).

In 1984/5 ICARDA established a program to train operators in the functions of the suction seed harvester, and to demonstrate seed bed preparation and sowing techniques necessary for a seed production unit. They had some success in this and their own seed production unit continues to this day, but the suction harvester proved a problem.

In 1987 delegates at the expert consultation at Sidi Thebet reported their attempts to produce commercial quantities of seed. In Tunisia the government had subsidised farmers to sow and harvest medic seed. Poor seed bed preparation (with conventional disc and plough) had caused yields to be low and inefficient harvesting had reduced them further. Farmers were no longer interested in producing seed (FAO, 1987, pp. 1–2). In Morocco a national centre had been set up. The same problems occurred – poor seed bed preparation and inefficient harvesting. The Moroccans were determined to continue the program (FAO, 1987, pp. 10–11). There was no suggestion that 'adapted varieties' would solve the problem – wisely there was no attempt to produce pure varieties of seed – but delegates wanted help to achieve better seed bed preparation and training in how to use the suction seed harvester.

A suggestion that farmers should collect their own seed and thresh and scarify by hand before using it to extend their fields was dismissed with scorn as 'primitive' (FAO, 1987, pp. 20–2).

While some considered the Australian seed expensive, others felt the cost was not high. Large quantities of vetch seed and huge quantities of feed grain for livestock, for instance, are routinely imported at considerable cost. The existing state monopolies that imported and sold-on seed were concerned that any medic seed industry should come under their control and this was a factor in the desire for a local seed industry.

Attempts by local farmers to produce profitable seed crops to this stage had failed because of poor seed bed preparation and difficulties with harvesting. If the farmer was to make a profit he would need a scarifier or combine seeder to prepare the level seed bed necessary for even germination and then he would need access to an efficient machine for the harvesting. A further consideration would be whether seed production would pay more than grazing his sheep on medic pasture. Given the generally high returns from sheep, the price he received for his seed would have to be very high.

In 1989 ICARDA announced that in view of the results from its seed production training program it intended to search for more appropriate seed harvesting technologies to replace the 'expensive suction machine', which they had decided was not suitable for the region (ICARDA, 1993).

In 1990, the Australian farmer employed on the project in Jordan successfully produced commercial quantities of medic seed using a combine seeder to prepare the seed bed and a suction harvester to harvest and thresh the seed. The seed was used by the project for its trials and its distribution to farmers of a forage mix of vetch and medic seed.

In 1993 ICARDA's Network Newsletter reported that farmers were being encouraged to collect their own seed with a hand-operated mechanical sweeper that picked up pods in the fields. In addition, ICARDA was seeking funds to manufacture a small medic pod thresher (ICARDA Newsletter, 1993, pp. 21–3).

Surprisingly the same Newsletter reported that 'until now there has been no indigenous medic seed production in WANA (West Asia and North Africa)' and refers to imported Australian seed as 'expensive and not always well adapted, especially in areas with cold winters'. ICARDA's next step, now that

adapted medics have been identified by researchers in the national programs, . . . is to multiply the seed up to useable quantities. Attention will be paid to methods of preparing fields for harvest, demonstrating the machines with other species, cleaning the seed pods and storing seed (ICARDA Newsletter, 1993, p. 22).

It was unfortunate that the work in Australia on refining the harvesting and threshing process remained unknown to the ICARDA group.

In 1988 a group of Seedco farmers in Australia had worked with a manufacturer to separate out the functions of harvesting and threshing. The result was a pod suction harvester that was much simpler than the existing combination harvester/thresher. This separation of function had the advantage of giving the farmer the choice of threshing the pods for seed or using the pods whole as a source of stored fodder for sheep that could be kept for a drought or sold to other livestock owners. A separate thresher had another advantage. Because the threshing could be done at a chosen time, it could be hired rather than bought as part of the original capital investment. The Ridgeway hand or machine-driven seed thresher using a cone invented by an Australian farmer had been made known to FAO and other institutions in the mid-1970s, yet it languished in Australia. This is particularly successful with medic pods as it rubs the pod rather than hitting them like the conventional thresher. The pod is highly resistant to being hit but it breaks up quite quickly and easily when rubbed. All these machines were known to various international institutions but none were sufficiently interested in them to use project funds to purchase and test prototypes on farms in the region.

After more than a decade of institutional concern with the establishment of a seed industry, programs to provide a centre from which to disseminate the techniques of seed production seemed to have gone round in a circle.

The FAO conference and the ICARDA outposts throughout the region have remained strangely untouched by what has been taking place elsewhere and particularly in Australia. One could expect that as Australia is the country where farmers discovered and successfully use a medic/cereal rotation that their innovations would be the focus of institutions in the Northern Hemisphere interested in the way in which medic can be used in a dryland farming zone. The lack of awareness of what is happening elsewhere – farmers collecting their own pods and using them to extend pastures and the seed production units in Morocco, Tunisia, Algeria and Jordan or the harvesting of pods for stored fodder and the harvesting machinery refinements in Australia – indicate that the institutional information network that should keep up to date has failed. Not until 1994 did ICARDA suggest that it may be possible to sow pods rather than seed, to establish pastures (ICARDA Newsletter, 1994).

Twenty years after the first seed production unit was set up, apart from the small quantities of seed produced at ICARDA, the region is still dependent on Australia for bulk supplies of seed. The storing of pods for drought reserves is not under consideration. Poor seed bed preparation that impedes yield and harvesting efficiency, identified throughout the region as a major barrier to a successful seed industry, continues in spite of the demonstration in Jordan that a scarifier can overcome this.

(f) Extension programs

During the 1980s and increasingly in the 1990s there were many calls from staff within various institutions for extension programs to help solve the problem of what they liked to call 'the failure of the ley farming system'. The question of what messages needed to be taken to the farmers remained obscure, let alone who was best qualified to take it to them. There was much talk of 'appropriate tillage' and 'grazing management' but when it came to the specifics it was difficult to define what the technical and scientific experts within the institutions meant.

As we have seen above, shallow tillage using a scarifier and the incorporation of the scarifier into a combine seeder seemed an obvious answer to the problem of 'inappropriate tillage' yet almost no work was done by the institutions in adopting the scarifier on their own farms or encouraging its purchase and distribution to farmers on theirs. Without the scarifier on the farm there was little point in carrying out extension programs to explain the benefits of shallow cultivation.

Grazing management was constantly criticised. The fear of incursions of nomadic diminished in importance as time went by – perhaps because it had not been a great problem after all. The faith in fences as a form of grazing management became less strong as the role of shepherds became more widely understood and perimeter fences proved of less importance than agreement between tribes, villages or individual farmers. The specific matter of managing grazing on medic pastures, either as permanent pasture or in rotation with cereals, was barely understood, let alone practised by the institutions. They were more comfortable with medic being sown and grazed as a forage crop. If they did carry out grazing trials of a pasture it was done, not on farms but on trial plots or fenced off mini-fields in an artificial manner that had little relevance to the farming system in which it was intended to operate. What was the point of setting up extension programs for farmers if the pasture phase was not fully understood and experienced by the technical staff who would instruct the extension agent?

Dr Cocks of ICARDA believed that farmers should get their information from 'consistent access to advisers and scientists to help them resolve problems as they arise' (Cocks, 1985, p. 17).

A colleague, Dr Nygaard of ICARDA, warned that introducing a new farming system could be dangerous if 'aspects of the technology [are not] understood by farmers or by those directly advising them' (Nygaard, 1980).

There were, of course, the expert medic farmers who had trained the farmers in Libya and in Jordan, and who had been employed as tractor drivers in Iraq and Jordan, and demonstrators and men-of-all work for two brief periods in Morocco. They not only knew the medic farming system inside out, but they also knew the constraints under which the Arab farmers were working.

Dr Cocks agreed that the Seedco farmers had done a good job in Libya but he reflected a common opinion within the institutions that

Australian farmers are no better at understanding [local conditions] than Australian scientists. Indeed, in many cases, through education and background, they are less able to adapt themselves to a completely new culture. In illustration I offer the example of frustration which Australian farmers have when trying to teach Arab farmers about machinery. They simply cannot understand that these people have no background in operating machinery and come to the conclusion that they are stupid. The idea that the machinery itself may be inappropriate is not a welcome idea for most Australian farmers. There is no universal brotherhood of farmers and the sooner we get rid of this idea the better (Springborg, 1985, p. 26).

The evidence, including their own accounts, of the Seedco farmers' involvement in Libya, Jordan and Morocco did not support this opinion, but if they were not to be used, how effective had been the extension undertaken by the institutions?

Technical experts who had worked on the South Australian demonstration farm at El Marj paid tribute to 'some extension' undertaken by the Seedco farmers and representatives of machinery companies supplying Australian machinery, but they wanted an extension service staffed by 'graduates receiving training in the principles of dryland farming' and they regretted that no extension programs had been undertaken to induce farmers to use herbicides for weed control and to make hay (S.A. Government, 1980, **2**, pp. 147–50).

In 1987 the FAO expert consultation in Sidi Thebet called for better extension programs and delegates were told that FAO was paying for the production of some audio-visual kits and a manual on the establishment and management of medic pastures that would explain the system and its operations to farmers, extension agents and technicians.

At the 1989 ICARDA workshop at Perugia the matter of extension programs was much discussed. It was agreed that

- a network was important as was training in extension methodology and ley farming practises for extension agents,
- 'specialists in ley farming' should be trained to provide back-up for extension agents,
- extension workers should receive more field support and extension programs should be closely related to research and both aspects must involve farmers,
- and 'despite some of the outstanding problems, work on ley farming should proceed to an extension phase . . . a high priority task should be to seek funding for a coordinated training and extension project to take ley farming to farmers . . . (taking full account) . . . of experienced gained in the earlier work' (ICARDA, 1993, p. 295).

The Australian team in Jordan had been delighted to attract $US500 000 to carry out extension programs including the preparation of material, but they were assisting farmers to undertake a forage program not a medic/cereal rotation.

The Moroccan delegates to the workshop were not so worried about extension programs. They had, in any case, dismissed the medic farming system as fitting poorly 'into the predominant socioeconomic conditions of the Moroccan agricultural society' and had recommended that in future medic should only be considered for 'bigger farms in the semi-arid zone'. During 'Operation ley farming' they claimed 'considerable efforts were made to assist extension staff and farmers by training, field days and demonstrations' (ICARDA, 1993, p. 36). The major extension program undertaken was the effort of the two Australian farmers who were employed for a number of weeks each autumn to come in and demonstrate the sowing of the pasture. They were required to work with everything from ploughs and discs to tree trunks dragged behind tractors to sow the seed, and also to instruct farmers, technicians and extension workers in the technicalities and management of a medic/cereal rotation during the few hours they spent on the sites. Their reports of their programs are a monument to their energy, good nature and good will, but a testament to the lack of understanding of the level of extension effort needed to insert a new farming system into a community dependent upon a technocracy educated firmly in the old one.

Between 1988 and 1990 the International Fund for Agricultural Development (IFAD) incorporated a brief period of demonstration and training for extension agents in projects in Tunisia, Algeria and Morocco.

In 1988 ACSAD funded for the Algerian Government a study into the resources needed for a national program of demonstration and training in the operations and management of the medic farming system to provide technicians and extension agents with basic expertise. The results of these will be discussed in the next chapter.

In 1989 FAO published a manual on management of medic pastures that collected together research results from Australia and the region to that time and detailed the resources needed for a medic/cereal rotation on farms and the establishment of permanent pasture on rangeland and how to use them. A series of four audio-visual kits with an accompanying booklet was also produced to illustrate the farm operations and management of the system. Translations in French and Arabic were made, but the distribution of these extension aids was not pursued enthusiastically (FAO, 1989).

By the mid-1990s the need for an 'extension effort' was rapidly replacing the search for an 'adapted variety' as yet another hare to explain what was still referred to as 'a failure of the system in the region'.

A working model that illustrates the variations possible with the system year after year to the farmer must form the kernel of useful extension programs. Yet no institution in the region had demonstrated that it could on its own stations run a continuous medic/farming system in which scarifiers prepare the shallow seed bed and control the weeds and in which a permanent flock of sheep relies for the major part of its nutrition on medic pastures and cereal stubble. The small farms in the region that had been touched by programs from institutions and research centres had varying success but none on such a scale that the system could be said to be firmly in place. Most either left the programs of their own volition or were left without support after a short period.

Without developing their own expertise in the operation of the system, how can the institutions produce extension programs that will assist farmers to adopt the system? The Seedco farmers in Libya provided a prototype for a decade of what is needed but the institutions are not prepared to accept its validity.

The winding down of institutional effort in the 1990s

However incomplete or short term the projects and programs were, the medic/cereal system continued to provide proof of its suitability to the region and its superiority over other suggested replacements for fallow. So why was it being flagged by the institutions as a 'failure in the region' to such an extent that programs to support its adoption were being wound down?

In 1989 Cocks and his colleagues from ICARDA presented the workshop in Perugia with a promising account of a sample of six small farms in Northern Syria on which they were attempting, with the farmers, to develop an indigenous medic farming system. Between 1985–8 the farmers had practised a rotation – several maintaining their pasture for two years instead of one – and the results had been good. The small locally used cultivator had not appeared to damage the medic seed bank although the seed bed preparation was not good and weeds remained a problem, profits from sheep had been good and cereal yields had improved. Yet in 1993 ICARDA was reporting dismal results from its WANA (West Asia and North African) program which was to have been based on the experience of these farms. Doolette's suggestion that confusion and misunderstanding about the system were common within the institutions was undoubtedly so.

In 1990 FAO published the proceedings of a workshop on Mediterranean pastures held at Bari. A paper on annual medic was printed as '*medicago sativa*'. The technical editors did not seem aware of the way in which the translation made nonsense of the content of the paper (FAO, 1990, p. 9). The manual on annual medic pasture published in 1989 was marred by poor editing that reversed a significant graph and cut out portions of important information, and some of the translations were confused.

ICARDA in its initial medic programs had promised well. Under Cocks' guidance it was prepared to attack some of the farming problems identified by various projects. Yet by the 1990s most of this early effort was fading away.

From 1990 ICARDA began to re-assert the emphasis of its programs into forage and grain legumes.

- It put large resources into persuading farmers to plant lentils and chick peas and little into persuading them to sow medic pastures. Major trials were taking place to test the amount of nitrogen returned to the soil by lentil and chick pea crops. The result indicated that nitrogen was greatly diminished by the need to harvest them when they were mature. The ICARDA pasture group in the mid-1980s had reported that medic pasture could fix 'up to 150 kg/ha of nitrogen' and have it available for the succeeding cereal crop.
- *Atriplex halimus* and *Salsola vermiculata* plants were being raised in the nursery, planted and watered for use in rangeland improvement. Nothing was being done to trial medic for the same purpose, but a new program was initiated to plant atriplex as a windbreak for barley crops where erosion was threatening barley production. Trial results on the station and nearby farms in the 1980s had confirmed the beneficial effects of medic pasture on eroded land and cereal yields, surely a much less expensive means of dealing with the problem. Under the heading 'Ley Farming' there was the announcement that the Tropical Agricultural Research Centre of Japan had provided US$300 000 for a satellite and balloon survey to assess the amount of vegetation in the rangeland.
- Grazing trials were being undertaken to test the nutritional value of cereal straw and it was found that ewes lost weight as the quantity of straw diminished. Thompson's trials in the 1980s with medic had showed that ewes gained weight on medic straw and pods.
- Narbon vetch, common chickling, subterranean vetch and woolly-pod vetch were being advocated as fallow replacements by ICARDA, yet farmers were still resisting their use. Grazing trials were undertaken in 1991 that showed that rotation of vetch with barley produced more animal feed than a continuous barley regime. The cost of manually collecting the vetch seed was posing 'severe limitations' to the adoption by farmers of the rotation. It was noted that medic pasture rotated with wheat and/or barley freed the farmer from the need to sow more seed and cultivate the land in the pasture phase but it does not appear to have

induced ICARDA to give medic pasture more priority in its program.
- There were still supporters of the moisture conserving aspect of bare fallow.
- The inoculation program for medics had been expanded to search for inoculants for food legumes. A small quantity of inoculated medic seed was being produced and distributed to other centres in the region and a germplasm program in Algeria had resulted in something over 300 lines of medics being identified as part of a general search for annual legumes.
- Medic seed had been suction harvested off 2 ha of the ICARDA farm site and the pulverised soil left behind after the operation enabled a measurement to be taken that showed that soil was removed by wind erosion.

There was no mention of the progress of the farms at Tah or the other farms encouraged by Cocks to try out a medic/cereal rotation although the WANA program (to encourage interest in medic pastures in the national research centres in the region) was continuing. The message from ICARDA in 1991 urged farmers to plant crops of Faba beans, chick peas and lentils, use more nitrogen fertiliser and take up improved varieties of wheat and barley if they wanted to increase their yields (ICARDA, 1991).

By 1993 the Australian governments had pulled their technical experts out of the region. They retreated to Australia and concerned themselves with extending what they had learnt about medic in the region to their own arid zones and to quantifying and refining the system used by Australian farmers (Squires & Tow, 1991).

The impression one has from the published reports is that by 1993 attempts to establish a medic farming system on farms in the region had dwindled away, its simplicities lost within the institutions and agencies as they diligently pursued narrow lanes that led nowhere. The farmers were losing the system by default as they continued to be denied access to the resources necessary to its successful operation.

The obtuse refusal of the institutions and agencies to face up to the clear needs of the farmers is apparent in a report detailing where resources went during the 1990s push by ICARDA to extend the use of a medic farming system by means of the WANA program. In Algeria the program was used for

- a trial to see if deep tillage was necessary,
- medic being sown and used as a forage crop with barley and oats,
- seed collection of a 'local' medic to add to the forage crop,
- trials to assess the response of nitrogen and superphosphate fertiliser to rainfall,
- trials of sheep management, breeding and nutrition,

- a survey to see if integrated livestock/crops were common (ICARDA Newsletter, 1993, p. 10).

The results confirmed that farmers continued to deep plough, that cereal yields were low and that no legume pastures were to be seen outside research centres, although 'a few were reported to exist on private farms'. The Chef du Cabinet of the Algerian Ministry of Agriculture (Benzaghou) had made a special plea to ICARDA to make this project 'significantly different' to the conventional one (ICARDA, 1993). He may well have been disappointed.

It is little wonder that the Director General of ICARDA (Nasratt Fadda) described the WANA program as a 'limited success'. This was due, in his opinion, to

the location-specific nature of various components of the system and the approach used to transfer the technology ... materials and practices used in developed countries were not adequately screened for their appropriateness in the recipient countries ... Algeria, Egypt, Iraq, Libya, Morocco, Syria and Tunisia. The fact that agricultural technology – including aspects of ley farming technology – is location-specific suggests that the International Centers alone are not equipped to undertake the task in every region of WANA. A partnership of all concerned is a prerequisite for success (ICARDA, 1993, p. 8).

It was, in effect, a confession of failure – but the reasons for it were not valid.

- The scarifier carried out shallow cultivation efficiently in Libya, Algeria, Iraq, Jordan and Australia.
- Medic is native to North Africa and the Near East.
- Grazing management is not beyond the wit of the Arab farmer.
- The Seedco farmers in Libya had succeeded due to a partnership between themselves and the Libyan farmers.

Would the institutions and agencies have had such limited success if they had established a partnership with expert medic farmers? Cocks talked of a partnership with the local Syrian farmers but not a partnership with expert medic farmers. The contrast with the results of the Seedco work in Libya is stark when put against the results of the ICARDA program in Algeria. The work done by Cocks and his colleagues at ICARDA in the mid-1980s in attempting to evolve a farming system on local Syrian farms using medic may have been slow and sometimes misguided, but it did at least conceive of the exploitation of medic as a farming system. Its weakness was in believing that the plant could be manipulated to overcome the tillage problem and the seasonal variations of frost and drought. By the 1990s medic pasture had become simply a minor component in the conventional list of research programs that told farmers nothing they did not know and rediscovered again and again that the old system in whatever guise is not going to solve the problems of erosion and declining yields.

Some reasons for failure

Throughout the forty years of endeavour to come to terms with a medic farming system institutional conferences, workshops and surveys have contained many reports of poorly defined research objectives that often produce ill-informed opinions about the way the system operates and why the attempts to have the farmers adopt it have failed.

Of course there are excuses to be made on behalf of those who have been working with medic in the region. The hierarchy of the research station or project creates a dependence on labourers to actually do the farm work and requires the technical expert to carry out various research trials, often ephemeral to the system he is interested in. This does place constraints on good management. Problems due to the absence of scarifiers and combine seeders, late arrival of seed, undesirable ploughing intrusions, stray sheep and so on have skewed results. The relatively short term of the trial or program or project is also a complicating factor. One should not overlook the rigidity of the initial plan – once the directives are given for a long term trial, or a five year project, for instance, it is difficult to adjust it in the light of new thoughts or unexpected occurrences. Too often the money has been spent, often on auto-harvesters to take off a crop that will never eventuate.

One can see the frustration of many technical experts *vis-à-vis* their difficulties in getting to grips with the practical management of the system. The seed collection and selection programs, the search for an 'adapted' variety and its multiplication provide a controllable, familiar, and comfortable refuge for those daunted by aspects of the system they feel uneasy about.

This does not seem a good enough reason to abandon the medic system and allow support for it to sink to a priority below grain legumes and forage crops. Many of these problems, after all, were faced and overcome by the expert farmers employed in Libya, Morocco and Jordan.

To call (as Nasratt Fadda did) for 'a partnership of all' because the institutions, after forty years of effort, have not been able on their own to persuade farmers that a medic/cereal farming system is a viable alternative to bare fallow poses the question – a partnership with whom? If the clear messages from the region's farmers are not heard and the experience and knowledge of expert medic farmers is to be excluded and their demonstration of a medic farming system within the region is to be ignored, to whom can the institutions turn?

11

On the farms in Tunisia

Introduction

The difficulty the institutions and agencies had in realising a viable model of the medic/cereal rotation and managing it successfully left the end-users, extension agents and the farmers, in an invidious situation. Farmers floundered not only because they had no scarifiers or combine seeders but also because of the confused messages about management that they were getting from the institutions. In spite of this many farmers persisted with medic and their governments and some agencies made efforts to help them.

The importance of management

On the farm the management of the pasture is the critical factor in success. Figure 11.1 shows the effect that management carried out by the farmer at the appropriate time has on the productivity of the system. Note that this chart shows in a simplified way the management of the medic system. The effects of each operation are shown in italics and then accumulated through the arrows and lines to the final yield of cereals and the regeneration of the medic pasture.

Provision of resources

In order to carry out effective management, farmers need adequate resources and governments within the region were well aware of these by the mid-1980s.

Within the region some changes in attitude towards medic had occurred. Tunisia's Agriculture Minister (Ben Osman), and many others, proudly claimed medics as indigenous species (FAO, 1987), although the suspicion of 'Australian cultivars' as cause of 'failure' was still commented upon. The

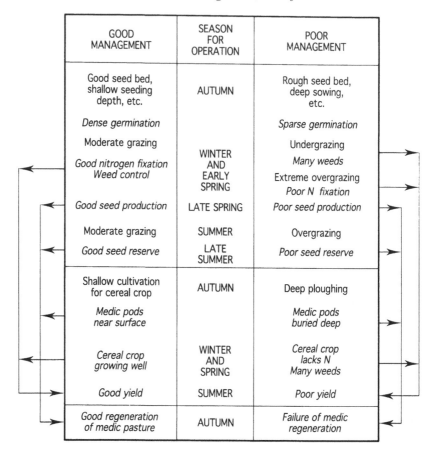

GOOD MANAGEMENT	SEASON FOR OPERATION	POOR MANAGEMENT
Good seed bed, shallow seeding depth, etc.	AUTUMN	Rough seed bed, deep sowing, etc.
Dense germination		*Sparse germination*
Moderate grazing		Undergrazing
Good nitrogen fixation *Weed control*	WINTER AND EARLY SPRING	*Many weeds* Extreme overgrazing *Poor N fixation*
Good seed production	LATE SPRING	*Poor seed production*
Moderate grazing	SUMMER	Overgrazing
Good seed reserve	LATE SUMMER	*Poor seed reserve*
Shallow cultivation for cereal crop	AUTUMN	Deep ploughing
Medic pods near surface		*Medic pods buried deep*
Cereal crop growing well	WINTER AND SPRING	*Cereal crop lacks N Many weeds*
Good yield	SUMMER	*Poor yield*
Good regeneration of medic pasture	AUTUMN	*Failure of medic regeneration*

Figure 11.1. A calendar of good and poor management of a medic rotation. (Source: B.A. Chatterton.)

identification of 'local cultivars' had taken a great leap forward and on most research stations throughout the region one saw trial plots of seeds from local cultivars. Often we were driven by enthusiastic technicians to a site (sometimes just a village common) specially to see a local cultivar flourishing undisturbed by deep ploughing. The problem of bulking up supplies of seed from these local cultivars was being tackled.

Funds for scarifiers and combine seeders, however, continued to be unobtainable due to decisions made in the opaque investment centres of agencies and ministries, although funds for purchases of equipment used for deep ploughing and for auto-harvesters remained as accessible as ever.

A common complaint from ministry officials and technicians involved in extension was that, having succeeded in growing medic successfully in the establishment phase, the continuation of the rotation proved difficult and they were frustrated at the lack of information available to assist them to advise farmers how to exploit the medic fully.

Requests continued to be made to the international agencies for assistance from expert medic farmers to provide this but to little effect. When the employment of expert farmers to advise and demonstrate was frustrated, other attempts were made to deal with the knowledge problem. The following case studies illustrate the way in which the Ministries of Agriculture and their associated institutions tried to develop a medic farming system for their own circumstances.

Medic on Tunisian farms – the early period

After the CIMYTT project finished in 1976 the Union of Cooperative Producers (UCP) that owned large farms in the Le Kef/Siliana governorates in north-west Tunisia continued to grow medic pasture with varying degrees of success. Frequently medic disappeared when a manager was replaced by a new one who was anti-medic. Some private farmers in the more marginal zone also persisted.

In 1979 we found quite good areas of grazing in places and many farmers who were pleased with the results both to their livestock and to their cereal yields. We seldom saw overgrazed pastures. Livestock owners were grazing the pastures quite well and some were exploiting the dry pods and straw which they used, together with the cereal stubble, during the summer months. On the research stations there appeared to be less effective grazing management and in one case we saw medic growing over one metre high because it had not been grazed at all.

But there were problems emerging with diminishing regeneration of pasture and increased weeds in cereals because farmers were forced to prepare the cereal seed bed with discs and chisels. They tried to use them so that they only penetrated the soil to a shallow depth but this was difficult and the cultivation achieved was far from perfect.

Cereal farmers nearer to Tunis in higher rainfall country were not so happy. They had tried to use the medic in rotation with their cereals but said that the subsequent increase in weeds after the pasture phase had caused their cereal yield to drop and they had lost interest in medic. They had used ploughs and discs to prepare the cereal seed beds adapting them to shallow depths. When weeds got too bad, they reverted to bare fallow.

Within the Ministry of Agriculture there was much enthusiasm about the potential of medic and strong support was being given to the continuing programs to introduce it to Tunisian farms. In spite of the awareness that the seed bed preparation for the cereal phase was causing a problem with regeneration of the pasture, there was no program to import or manufacture scarifiers in order to overcome it.

Medic on Tunisian farms – the mid-1980s

Between 1986 and 1988 we carried out a study of small cereal and livestock farms in the Le Kef and Siliana governorates where medic had been introduced as part of a large project designed to improve farm productivity. The project had begun in 1980.

It was initially proposed to plant 18 600 ha of medic on private farms varying in size from 5 ha to 20 ha. By 1985 a total of 3568 ha had been established. The cultivars Jemalong (*M. truncatula*), Harbinger (*M. littoralis*) and Paraggio (*M. truncatula*) (all from Australia) were used and planted at the rate of 15 kg/ha together with 100 kg/ha of superphosphate. The medic grew and produced well in the first year but regeneration after the cereal phase was often poor. Interviews with the farmers revealed that seed beds for cereals had been prepared using deep ploughs. The farmers and the technicians responsible for advising them about the new system knew that deep ploughing was death to medic regeneration, but no one had a scarifier. No demonstration of the economies available with shallow cultivation was made to farmers or technicians. Two large autoheaders had been purchased to harvest the cereal crops that were envisaged.

Deep ploughing was not the only problem. Farmers who were supplied with medic seed in the first instance did not all own sheep. Some who were allocated sheep did not get them in time for the spring grazing and so their medic pasture was unused. In the following year, the allocated sheep arrived on time, but it was a drought year and the pasture was poor and farmers had to buy in feed. This illustrates how an administrative blunder together with a seasonal hiccup can lead to a technical expert in the metropolitan institution deciding that medic is a failure.

By spring of 1986 many farmers were in debt and having difficulty repaying loans and the technicians were discouraged at the lack of progress. Nevertheless, the project continued to supply medic seed to farmers, albeit on a much smaller scale and with a very much smaller budget to carry out the necessary establishment operations. In the autumn of 1986 in one district, a little over 1 ha of medic was sown on each of 50 small farms. This time, all the farmers involved owned a small flock of sheep.

Germination of the medic was good, but growth was slow because of late planting due to the unavailability of the machinery from the government depot at the proper sowing time. Thus by November, when the spontaneous medic that grew on the roadside and on unploughed parcours in the same district was making good growth and could be grazed with impunity, the sown medic plants were still small and needed several more weeks before sheep could be introduced to the pasture. In 1986 a lack of information about the management of medic pastures was identified as a fundamental problem in the operation of the project and project funds were made available to try to rectify this. The first step was to interview farmers to see how they felt about medic.

Surprisingly many farmers were still keen to have medic.

Farmers' initiatives in response to lack of resources

The farmers explained why. In the first year when farmers had medic pasture but no sheep some of them had made hay and recouped some of their costs. When the sheep arrived after they had made their hay, they kept the sheep in anticipation of good medic the following year but there was a drought and regeneration was poor, so they sold the sheep. They had been impressed by the amount of feed the medic had produced. Those who in 1986 began again with a newly planted hectare or so of medic were keen to do well out of it. Although the official project report claimed that medic from the previous years had not regenerated, visits to the farms revealed that, in fact, some had regenerated (one field in particular had been regenerating well for four years) and farmers were using it either to provide grazing in spring or cutting it for hay. One or two cut hay during the flowering period and did subsequently lose their medic, but others cut before flowering or after the pods had dropped and their pasture regenerated quite well.

Grazing management

The 1986 sowings of medic grew well. The next requirement was to graze it properly. No one seemed to have anticipated the abundant pasture that became available. Most farmers only had small flocks and these were not enough to graze the medic hard enough to avoid weed infestation. This was exacerbated by advice given by technicians. Their experience was with forage crops and they told the farmers to let the medic grow 50–75 cm in height and the allow the sheep to eat it off in one grazing operation. They did not understand the difference between grazing forage crops and grazing medic pasture. Medic grows better the more frequently it is grazed. It is the grazing that encourages the spread of the plant and the production of pods. It is also this frequent grazing (the utilisation of the growth potential of the plant) that enables medic to provide a large quantity of feed over a long period. Not to utilise this growth potential is to lose a great part of the economic benefit of medic pasture.

A few technicians had been impressed by the message that a good seed bank was important for future pasture regeneration and they advised farmers not to graze the medic at all in the first year in order to save the seed for the following regeneration. This was intended also to overcome some of the effect of deep ploughing and the loss of seed that was buried too deep to germinate. The result was that the first year's growth was wasted and weeds became a huge problem not only in the cereal phase but also in the subsequent pasture phase. Fortunately, few farmers took this advice as they did not have enough land simply to allow a crop to luxuriate untouched. Several farmers invited their neighbours in (as did the farmers in Tah in Syria) to graze when their own flocks could not use all the pasture available. Most charged a fee (decided between themselves) and were quite pleased with the transaction.

One large UCP farm (Ennajat) maintained a year round stocking rate on cultivar Paraggio (*M. truncatula*) equal to 4.7 sheep/ha. Farmers reported savings of between 40 and 181 TD/ha in purchases of grain and hay, but those who grazed less saved less. Those farms where grazing was heaviest had less weeds in the cereal and pasture phases and produced more medic pods from their pasture. The UCP farm had the largest reserves of medic pods and even at the end of summer when it was inspected, 700 sheep had been grazing 80 ha all through summer and had not been able to consume all the straw and pods and a great deal remained.

We had developed, with the help of Dr Higgs in South Australia, a simple, cheap device for measuring pods and seed during summer. It was demonstrated to technicians and farmers alike and it proved effective in giving them confidence to allow the sheep to graze the summer pasture.

Benefits to the farm family of grazing medic

The normal cost of feeding the six or so sheep kept by the farm family and their two or three cattle was high. Farm families did not (as many non-farmers believe) expect their animals to survive on weeds on the side of the road. Grain and hay were needed to keep the animals alive and producing milk and meat. The cereal the family grew was not only for themselves but for their animals and straw for winter reserves. They had to buy what they could not produce and that tended to be the greater part of the year's feed. It was clear from discussions with the farmers that the immediate benefit of the medic was the reduction or even elimination of the cost of purchased feedstuff. The next benefit was increased live weight gain in lambs. Increased stocking rates were something to consider later. This contrasts with the objectives frequently outlined by the institutional centres which tended to concentrate on stocking rates rather than reduced costs.

Farmers' reasons for choosing when to rotate

Many farmers and most technicians were unsure about whether to leave the medic for a second year to provide more pasture or whether to sow cereals. When farmers were questioned as to why they intended to plough and plant cereals on the pasture land in the second year, they could give no cogent reason. When cost and return comparisons were worked out in the field with them using their own figures they found that often it was a better proposition to keep their medic pasture for another year than to intervene with a cereal crop. (Remember that only part of the farm was sown to pasture in this first year, leaving the rest for the normal cereal crop.) Where several farmers had, of their own volition, left the medic pasture to regenerate uninterrupted by a cereal phase, they were content with the results even though at times the pasture looked to us to have a high weed content. They contrasted the grazing obtained from medic and volunteer weed pasture with the conventional fallow of their neighbours and were in no doubt that they had the advantage. The technicians agreed with the economics of the proposition of leaving the land in pasture for one or more years before intervening with a cereal crop, but privately confessed that they had believed that is was essential to have a cereal phase in order to have the medic regenerate and that they had advised the farmers of this – thus revealing one of the many dangers in advancing a simple but rigid 'ley farming' concept where cereals follow medic pasture with unrelieved monotony. The flexibility inherent in the medic/cereal system that allowed farmers to make many choices about land use depending on their needs and circumstances was difficult for the technicians to grasp.

What farm budgets revealed

A quick budget drawn up in the field on farms using the farmers' own account of their costs and returns to demonstrate the relative advantages of leaving the medic pasture for a second year instead of sowing a cereal crop proved an effective means of drawing the project technician and the project farmer together in arriving at a decision about whether or not to enter into a rotation.

In addition to the information about the relative farm costs and returns obtained in this way, official statistics were obtained from the Unit for Economic Studies within the Office des Céréales in Tunis, and prototype budgets were developed to show farmers the difference in the costs and returns of the old system they had been using and the new one using medic pastures.

Surprisingly the cost of production of the conventional system being advocated by the Ministry of Agriculture was not low as one would expect from rather primitive farming being carried out on the small holdings. The cost of hire of state owned machinery (which included the time to and from the depot), the cost of certified cereal seed (even though subsidised), and the cost of large quantities of nitrogen fertiliser were extremely high and appeared to be a major reason why farmers tended to rely on mules and wooden ploughs, and their own cereal seed, and did not use fertiliser. The erratic and often late delivery of seed and fertiliser was a factor in the low cereal yields and the organisational weaknesses of the state machinery depots worked against the farmer's desire to avail himself of mechanisation.

It helps explain the fact that many farmers with quite small holdings (sometimes as little as 5 ha) will go into debt to buy a tractor rather than rely on contract services. This matter of excessive cost of mechanisation and other resources needed for an agricultural crop are not often taken into account when sophisticated rotations are being touted as a means of increasing production on small- and medium-sized farms, yet they are a potent reason for low crop yields and also for the rejection by farmers of the new technology. All the scientific data to prove the optimum time for sowing cereals count for nothing if, when the rains come, the tractors and implements needed are being used by someone else or lying in the machinery centre waiting for repairs to be made.

Machinery acquisition

It was obvious that the farmers needed scarifiers for shallow cultivation in the cereal phase. As a second-best alternative, it was suggested that several be purchased and made available to the local machinery centres so that farmers could hire them when needed. The project itself should purchase one so that it could be used to prepare seed beds and sow the demonstration plots on the project farms.

It was also recommended that one or two experienced medic farmers should be employed to show the farmers how to carry out rapid shallow cultivation and later to assist the farmers and technicians to develop an understanding of summer and winter grazing of the pasture.

The recommendations were agreed but no purchases or employment followed.

One of the lesser known difficulties attending the introduction of new machinery into a developing country was illustrated at one seminar during which the relative merits of deep ploughing and shallow cultivation were contrasted. The seminar was attended not only by Tunisian technicians but also by foreign personnel from a development agency that was concentrating on the provision of farm machinery in the region. There was a heated exchange of words about the utility of scarifiers and combine seeders, the agency personnel believing that the advice that scarifiers would prepare a shallow seed bed efficiently was an attempt to make incursions into the market they had established for their equipment designed for deep ploughing. On a subsequent occasion, and apparently after some considered thought, the agency personnel accepted the disinterest of the advice and volunteered to work with the project to build a prototype of a scarifier and to investigate the possibility of local production of scarifiers.

Training and demonstration program for technicians

It was obvious that the technicians responsible for advising farmers about the operations of a medic farming system needed not only to know more about the system in general, but also to understand the way in which the farmer could best manage the system on his individual farm. A series of seasonal training sessions was carried out in which farmers and technicians were involved. These proved to be valuable also in assessing the degree of knowledge about medic management and exploitation present within both the technical and farming communities. The joining together of the technician and the farmer in the course helped sensitise the technician to the farmer's practical problems in trying to manage the medic. They became aware not only of the farming reasons for adopting or not adopting various operations basic to the system, but also of the advantages the flexibilities inherent in the system gave to farmers coping with the vicissitudes of dryland farming. They gained a great deal more confidence in their ability to advise and assist the farmers for whom they were responsible.

Training sessions

The sessions were conducted during three seasons (mid-winter, spring and summer) and operated in the following manner.

- During the first day, film (usually a selection of photographic slides) was shown of the management and operations required in order to use the medic effectively according to the season in which the session was being held.
- On the second day, group visits were made to the farms where the medic was growing and discussions took place with the farmer about the management in the previous year that accounted for the current condition of the pasture. The group and the farmer then discussed the operations that could be undertaken in order to make use of the pastures and what management was needed to correct any problems that were apparent.

  In addition to this group visit, a questionnaire was drawn up for each technician to take separately to the group of farmers for which he was responsible so that a statistical base could be obtained showing seeding rates used for cereals and medic, choice and amount of fertiliser, number of ploughings carried out, depth of sowing, yield, number of livestock, stored fodder provided, amount of medic substituted for this, costs of all operations and so on.
- On the third day, the questionnaires were used as the basis for a discussion about the management of medic pasture within the project generally, and to draw out from the technicians their views about the farm operations to be undertaken during the forthcoming season.

The demand for training and demonstration

Enthusiasm about these sessions grew and during the summer there were often more than forty technicians present at each session. This presented problems as the sessions were designed for no more than twenty and preferably twelve technicians. The excess came in part from another project nearby, which was trying to increase production on cereal and livestock farms in the region, and in part with some managers of UCP farms where medic had been first installed in the early 1970s and had met with mixed fortunes, as well as some technicians attached to the Office des Céréales. General seminars were also held for students at the Ecole Supérieure at Le Kef and the Centre pour Récyclage at Siliana where film of various aspects of the medic/cereal rotation was shown and long, vigorous question and answer sessions followed. The centre offered to cooperate with the project to establish a demonstration unit for grazing and shallow cultivation, but it is not known if this proceeded.

Institutional support for education about medic

There was no comprehensive component on medic cultivation and management within the current curriculum of either institution and, in fact, some of the French text books still used were teaching principles such as deep ploughing, application of nitrogen fertiliser and intensive feeding for livestock that were in conflict with the requirements of the management of medic pastures (Duthil, 1967).

Other textbooks devoted to Mediterranean forage crops have specific sections on medics and subterranean clover but only describe the botanical characteristics of the plant and quote CIMYTT on its broad potential to the region without providing information about its management or its role as the basis of an alternative farming system for the region (Lapeyronie, 1982).

There was no lack of enthusiasm among staff of the Tunisian institutions for a comprehensive component of medic studies to be incorporated into the courses available for agricultural students, but there was no way in which they themselves could initiate such a course. Text books and other types of teaching material specific to medic farming were needed, but because the local institutions could not make autonomous purchases of them from within their current resources and the centralised purchasing agency for textbooks did not appear to be aware of the need, none were likely to become available.

Local cultivars

On the other hand it was simple for one scientist attached to the centre to get permission to carry out research into local cultivars of medic and he had identified and multiplied small amounts of several varieties. There was no lack of local cultivars available and many were observed growing on roadsides and on unploughed parcours (rough grazing land adjacent to cultivated fields). Several farmers were paying children to gather medic pods from these locations which they sowed on their farms in order to establish pasture.

Rehabilitation of the rangeland

A review was made of attempts being made to regenerate a large area of rangeland and parcours. As usual, the planting of fodder shrubs (mainly species of acacia in this case) and the exclusion of stock were the means adopted. A great deal of spontaneous local medic was seen and it was recommended that superphosphate in varying quantities should be applied to this to see what result could be obtained. If this did improve the existing medic it was recommended that broadcasting of suitable medic seed and phosphate fertiliser could be undertaken in order to increase the natural growth of the pasture and, if successful, a grazing regime initiated as soon as possible. The large range of medic varieties observed in the region, their ability to quickly dominate introduced species and their response to fertiliser on farms in the region indicated that such a program could rapidly elevate the rangeland and parcours to a reasonable grazing resource.

The aftermath

At the end of the 1980s there was a group of technicians with a good understanding of the role medic could play in improving the farms of Le Kef and Siliana. There were a number of farmers who had progressed quite a way in exploiting the medic on their own farms for economic gain. Unfortunately when the term of the project ended, many of the technicians dispersed to other centres where there were no projects to introduce medic onto farms or rangeland, the farmers were left without scarifiers and, except for those few farmers who will continue to puzzle things out for themselves, the thrust towards a better alternative farming system collapsed.

Saouaf farm

Concurrent with the projects described above, a livestock production centre at Saouaf under the direction of the Office de L'Elévage et des Pâturages in Tunis had been set up to use medic to boost production in the hope that this would provide a prototype for large farms in the marginal zone.

In the initial stage of this project (begun in 1983) medic was sown over all the 2000 ha site in order to produce sufficient feed to increase the flock of 500 sheep and also to produce seed for future demand. This farm was inspected on three separate occasions during 1986 and 1987 and discussions held with the farm manager.

The first sowing was successful and medic was flourishing in fields and particularly on parcours, where it benefited from dressings of superphosphate. The medic seed production area was not so successful as the field used was very stony and the undulations caused by the deep ploughing used to prepare the seed bed meant that the imported suction harvester was only capable of harvesting about 800 kg of seed from the total area. This was a yield of less than 100 kg/ha and was far below profitable levels, but did supply sufficient seed to replant medic pastures on the farm itself.

It was apparent to us that the farm was being run on the basis of two systems. That is, the medic pastures were certainly providing excellent grazing for animals in late winter and spring and summer, but the pastures were being ploughed up during the cereal phase using discs set too deep to allow successful regeneration. During this phase nitrogen fertiliser (100 kg/ha) was applied to cereal crops. The cereal was mainly barley for grazing and for hay and some grain storage for drought reserves. The medic was re-sown the following year. Where it was not re-sown as a matter of course, the regeneration gradually diminished over the succeeding years until it became necessary to re-sow. (This reflected the experience of the UCP farms where, because there were no scarifiers, deep ploughing was carried out for the cereal phase and carefully grazed and nurtured medic pastures were slowly but inevitably depleted until re-sowing became necessary.)

In spite of this management regime, the medic was flourishing. In winter and spring it grew abundantly and flowered profusely, so that in the summer vast quantities of pods and straw were piled high in drifts against boundary fences. The original imported cultivars had long become dominated by local varieties. By 1986 the farm had increased the flock from 500 sheep up to 20 000 for short periods in good seasons and the numbers rarely went below 5000 even in severe drought.

Development of the parcours

The parcours on the centre was being treated as rangeland and advice on how to regenerate it came from rangeland experts. The result was that atriplex had been planted on low lying ground that collected the rainfall in puddles and was too wet for the plant to thrive. Perennial acacia, to be cut and fed to sheep in summer during droughts, was also being sown. It required watering from carts and many plants died in summer. Perennial medicago bushes (*M. arborea*) were being planted as well.

During the inspection in March 1987 the following was observed. A tractor and trailer fitted with a mechanical planter was cutting through a thick ground cover of spontaneously occurring medic pasture in order to plant forage bushes (in this case *M. arborea*) that were then being watered from a tank being pulled by another tractor. We asked why the spontaneous medic pasture (thick and abundant and capable of supporting livestock through both summer and winter) was being destroyed in order to try to establish a perennial bush that needed costly care to establish and maintain. The reply was that it was 'policy'.

In the following July there were huge quantities of dry pods and straw of the naturally regenerating spontaneous medic lying on the ground or banked up against ridges of soil or fences available for grazing, while the costly and labour intensive plantings of perennial shrubs were already wilting and in need of regular carted water for their survival. Funds, however, were available for the implantation of fodder shrubs and trees on the rangeland and the advice from international technical centres was that these were best.

Staff training and its effect on management

The major problem with the production unit was that the on-farm managers had not been trained in, or had any experience of, the management of a medic-based grazing unit. The personnel involved were graduates of institutes that taught the system of sowing barley on deep ploughed seed beds and fertilising it with nitrogen fertiliser. Without a countervailing education they could only respond to the medic as a valuable forage crop to be sown on formerly fallow land. The contribution of soil nitrogen to the cereal crop from the medic pasture was ignored, and they continued to apply bags of nitrogen to the barley and the waste went unremarked. Although they grazed the parcours and even broadcast superphosphate on it occasionally, they did not draw from this any lesson about its capacity to regenerate without cultivation, or its value as a cheaper and more productive alternative to forage shrubs in such marginal country.

They did, however, graze the medic in a relatively sophisticated manner – that is, they did not use it as a forage crop. Sheep were kept on it continually wherever it was available on the centre. They conscientiously re-sowed the medic in fields where deep ploughing impaired the natural regeneration, and they produced medic seed and continued to harvest it by mechanical means in spite of the fact that the operation was time consuming.

Certainly the production from the farm was very high – almost astronomical in contrast to neighbouring farms where pasture was in the final stages of degradation, most animal feed had to be bought, and cereal yields were sparse and well below 1 tonne/ha. Nonetheless, it was a wasteful regime and while the medic continued to thrive in spite of the management rather than because of it, the benefits of the medic were not exploited to the full. The hope that the farm would provide a model for other private or cooperative farms in the zone was not achieved (Chatterton & Chatterton, 1986–7).

Conclusion

The Tunisian Ministry of Agriculture strongly supported the use of medic pastures to increase productivity on state, cooperative and private farms. In spite of this it proved impossible to provide the farms with scarifiers or expert medic advisers on long term contract. The Ministry supported the idea of training local technicians to understand the management of the medic system so that an extension service would be available to potential medic farmers, but were unable to provide funds themselves to continue courses when agency funds ran out. Without these resources, farmers were left to muddle on as best they could.

On the large farm at Saouaf, the exploitation of medic succeeded in boosting production but the cost involved in operating the dual and contradictory systems that the management used for the farm left neighbouring landholders both unconvinced and unable to follow suit.

12
On the farms in Algeria

Introduction

Following the completion of the FAO supported project in the mid-1970s, the Algerian interest in medic continued, but on a less formal basis. Only one cooperative, the Domaine Chouhada, at Khemis Miliana, out of the many that were sown to medic by the FAO project continued a rotation on part of the farm. We visited it a number of times and inspected the rotation with the farmer who was the resident farm manager. He was able to use a scarifier that had been given to the Algerian Government by an Australian manufacturer and so had no difficulty with regeneration of the pasture. He sold the cereal stubble elsewhere on the farm to nomad flockowners and had no difficulty in reserving the medic residues for his own flock. He did have problems with weeds in his cereals after pasture in spite of good grazing but he believed this was due to the increased fertility of the soil and he was able to develop a spraying program that was successful in controlling the weeds. His cereal yields increased after using medic pasture in place of fallow.

The research station at El Khroub had medic pasture which had also been sown in the 1970s and regenerated well. A scarifier and combine seeder were available on the centre for use in the cereal phase. Grazing management suffered from the feeding of the mandatory 200 g of grain per head to sheep each day. This caused some technicians to suspect that sheep did not really like grazing medic.

Medic on Algerian farms in the 1980s

In 1986 the Algerian Ministry of Agriculture decided to initiate a project to sow medic on ITGC stations, on nearby state farms, and on private farms in the cereal zones of Algeria. The Wilyas of Constantine, Mila, Guelma, Annaba, Tiaret, Sidi bel Abess, Tlemcen, Mendes and Chlef (formerly El Asnam) were involved. The ITGC already had on each research station a program for the identification and selection of medics – testing both the readily available commercial cultivars and locally growing varieties.

An integrated program was proposed – research into medic cultivars, broadscale plantings on representative farms with research personnel working hand in hand with local agronomists to prepare the programs, and extension agents to supervise farm programs and instruct farmers in the establishment and exploitation techniques. Various international institutions were asked to assist and we were employed to research and advise on a national plan for extension and training to underpin the program. In the course of the study most sites planted to medic were visited, extension agents and technical staff were interviewed, their work (that of setting up demonstrations on suitable farms) observed, and farmers were interviewed about their farm programs.

Cultivars and the climate

By this time there was a much larger range of cultivars available commercially than in the 1970s, but there remained a tendency to sow only one cultivar of medic in each field, which meant that if damage occurred due to frost, losses were unnecessarily high. For instance, while the Australian cultivar Paraponto (*M. rugosa*) did very well on the cold, high plateau, and Saphi, Borung and Paraggio (*M. truncatula*) did well elsewhere in frosty conditions, a previously well established area of Jemalong (*M. truncatula*) on a farm at Sidi bel Abess (a not particularly cold climate) was wiped out by frost. Although the cereals benefiting from the previous medic pasture were excellent and the farmer was very happy with the improved fertility that followed, he was faced with the probability of having to reseed the area that had been killed. Trial plots of the Syrian cultivar of *M. rigidula* on the nearby Sidi bel Abess station were also badly damaged by frost.

On several stations, technicians had established trial plots of an Algerian ecotype (*M. siliaris*), that was among varieties of native medic pasture we frequently saw (made up of a number of ecotypes – sometimes three, sometimes up to six) growing spontaneously on roadsides and on parcours. We also noted another ecotype (an Algerian *M. polymorpha*) that was also common in this spontaneous pasture. These various local varieties were identified in a paper on Algerian medics published in 1989 (Adem, 1989).

This common mixed pasture was also discovered on a private farm in the Tiaret region where the farmer told us it had been regenerating and providing pasture for his sheep for thirty years. He had sown occasional barley crops without causing any diminution in the productivity of the subsequent pasture. We also found a large untouched area of medic pasture on the ITGC station 'El Zachariah' at Tiaret, where for thirty years it had provided grazing for the previous private owner's flocks. The local ecotypes, unharmed by cold and frost and growing abundantly, were well able to sustain heavy grazing. The lesson that a mixture of cultivars rather than one should be sown in the establishment phase had only been learnt in part. Where mixtures were sown, for example in the Guelma area, the resistance to severe cold was uniformly good.

Sowing program and establishment phase

ICARDA was asked to advise on the cultivars and sowing rates for the project and they recommended that commonly available commercial cultivars be sown at a rate of 30 kg/ha. As part of our research to assess the result of this we used a questionnaire to find out what rates farmers involved in the project had used on twentyseven private farms. Thirteen sowed more than 20 kg/ha, twelve sowed between 12 and 20 kg/ha and two sowed about 10 kg/ha. The high sowing rate was recommended to avoid losses caused by rough seed beds but, in fact, the germination and plant cover was not significantly different between high and low rates. One sowing of 45 kg/ha on the station at Constantine failed completely. The 45 kg/ha rate appeared to have failed due to the deep ploughing of the site immediately prior to planting as on a nearby site the same seed had been used at 10 kg/ha, the seed bed preparation had been shallow and germination had been adequate. On the station at Tiaret and on several neighbouring farms, where an Australian-trained technician used only two scratch type cultivations to prepare a shallow seed bed before sowing 10 kg/ha, the germination and plant cover was very good.

The Algerian Ministry paid for the cost of the seed used in most cases. Had farmers followed the advice to sow more than 30 kg/ha this would have built an unnecessary cost factor into their subsequent farming system. The absence of a commercial medic seed industry within Algeria would (if this particularly heavy sowing rate were adopted) also require an even heavier dependency on imported seed once larger scale plantings were undertaken.

There was the usual competition for machinery, and so although rain came early in October, most of the medic was not sown until late in November and several fields were not sown until 5 December. Had scarifiers and/or combine seeders been available this would have been avoided.

Grazing management on farms

Many farmers grazed the medic as one would a forage crop. That is, they allowed it to become rank, then put the flock in to eat it down, then took the flock out again. They were very pleased with the result and many were precise about the savings they had made from not having to buy in so much outside fodder. There was (with few exceptions) no conception of summer grazing of dry pods and straw and all the summer production was wasted, except for the seed that was available for regeneration in the autumn.

Some farmers recorded stocking rates of six ewes plus lambs per hectare during a 'spring flush' of March/April, one carried twenty ewes per hectare in March 1987. One farmer grazed his sheep continuously from January until the cereal stubble was available in July. Some farmers sold their normal stocks of hay (made from meadow grass and cereal straw) for ready cash because they were not needed. On the stations and the government pilot farms, however, it was difficult to assess the value of the medic pasture as regulations decreed that all sheep should receive 200 g of concentrate and a certain amount of hay each day and this continued whether medic was available or not. At Guelma (under the supervision of a technician who had spent a year working on a South Australian medic seed production farm) grazing began at the end of January and was continuous and continuing. This almost weed free, good quality pasture had been regenerating for several years and carried a good quantity of sheep – on average, about six ewes per hectare plus lambs during winter and spring.

The technician in charge was also successfully harvesting medic seed from the pastures on the station, using it to extend the area of medic pasture. He was about to undertake a seed production course at ICARDA.

At Guelma and Sidi bel Abbess, where the technicians in charge were quick to see the advantages of replacing the standard purchases of straw and grain with medic pasture, the farms around were benefiting from their advice. On other stations and farms either the farmer was waiting for sheep to be made available by the government, or for advice from his technical adviser that he might begin his grazing program. These medic pastures were becoming rank and weeds were beginning to dominate. Some technicians had advised farmers to plant their medic pasture near to the bergeries (sheep sheds) to make the pasture more readily accessible to the sheep and also to cut down the time needed for travel to and from the field so that the maximum grazing could be achieved during the day.

In general, farmers undergrazed the medic because they did not have enough sheep to exploit the pasture to its full potential and they lacked instruction in how to manage continuous and controlled grazing throughout the year. Had the establishment sowing been related to the number of sheep the farmer had in his flock, the problem of undergrazing would have been avoided.

Shallow cultivation and seeding during the cereal phase

Although the FAO/MARA project in the mid-1970s had demonstrated that unless shallow cultivation was incorporated into a medic/cereal program the regeneration of the medic pastures could not be sustained, all farms involved in the new program continued to deep plough.

Investigation revealed that the ploughing was done with a mouldboard plough, followed by discs, harrows, rollers and more discing, harrowing and rolling. Most farmers had on their farms a light tyned scarifier made in the machinery factory at Sidi bel Abess, but it was not considered useful for primary cultivation, and it is doubtful whether the pressure on the tynes was sufficient to penetrate hard ground.

The majority of farmers interviewed had taken three months to prepare their seed beds, making on average seven passages over the ground. Several made nine passages, and only on two farms was there a simple two passage preparation. All fertiliser was put out separately and usually by hand. Only eleven farmers applied fertiliser. They used triple superphosphate and rates varied from 100 to 200 kg/ha. Seeding was done using either French and Danish machines fitted with discs, or by hand.

The Algerian Government had, following the recommendation from the FAO/Algerian project final report of 1976, investigated the Australian scarifier/combine seeder and in 1984 a number were imported. Two of these were seen during the course of the present study. One machine had not been used, and the other had its chain broken due to it being loaded back to front and fertiliser left in the box to harden. The Algerians had (when ordering the machines) requested the manufacturers to send a mechanic to Algeria to demonstrate its use and maintenance requirements, but that request was rejected. Although instructions were printed on the machine in French they were not sufficient to enable anyone unfamiliar with the principles of the combine seeder to operate it. In 1990 we had an opportunity to physically demonstrate the machine to staff and operators at the 'El Zachariah' station. They were impressed with its efficiency (particularly the way it opened up extremely hard ground with tynes) but seemed to believe that a single demonstration and a few words with the operator afterwards was enough training for the machine to become part of the station's working plant.

Regeneration of medic pastures

Regeneration after the cereal crop was patchy on most farms. The continuing use of ploughs, discs and chisels for cereal seed bed preparation produced the usual slow but continual decline in the seed reserve and we could see that the pasture would require re-sowing in the near future.

One of the liabilities with re-sowing is that some productivity is lost. The difference between the medic newly sown and the medic that was regenerating naturally was quite noticeable. Regenerated medic, even in the colder zones, was more advanced and well able to stand grazing some weeks earlier than newly sown medic. The importance of achieving regeneration to ensure optimum production was clear.

Unnecessary cultivation of pasture land

It became obvious that several misconceptions had taken root in Algeria about the rotation.

The message inculcated by visiting Australian technical experts that medic was part of a 'ley farming system' had convinced many technicians that a cereal crop *must* follow a medic pasture. They found it difficult to understand that a cereal crop was an option and that medic pasture could be left for a couple of years if this was profitable for the farmer.

Many Algerian technicians are also convinced that the ground on which the medic pods rest must be cultivated if the seed in the pods is to germinate following the autumn rains. Careful probing of this belief revealed that the education received at college or university (reinforced by booklets read and workshops attended during working life) leaves the technician with the certainty that all germination depends on cultivation of the ground and that the soil must be disturbed to allow in necessary air and water for growth. One can point to the thriving native medic at the side of the road where the soil has not been disturbed at all and contrast it with the medic struggling to survive because the ground on which it lies has been ploughed. This demonstration that the ploughing is not necessary to the plant's germination and growth is graphic, but the constraint of the technician's education is still very strong, and time and again the technicians expressed their personal difficulties at being put in the position of giving farmers advice on techniques they themselves had not seen and of whose results they themselves were not sure. Here again the results obtained during the FAO project in Algeria in the early 1970s seemed not to have filtered through into the general education being given to these technicians.

The dilemma of the changed role of the farm adviser

In 1987 the Algerian Government took the decision to privatise most of the large state and cooperative farms by dividing them up into units of 100–200 ha, plus a flock of sheep of about 100 animals, and allocating each unit to a groupement of families. It was then left to them to make the farms economically viable. The new groupements lacked the abundant resources of the old state and cooperative farms, and as most of the farmers involved had previously been peons or low level managers on the old farms they had no accumulated capital with which to equip and run their new enterprises. The technicians and extension agents involved in the ITGC program were sensitive to this and most saw the logic of replacing the old expensive system of fallow, nitrogen fertiliser and grain and hay with the cheaper medic/cereal rotation.

The core group of Algerian technicians chosen to lead the medic program all had some expertise in the establishment of medic pastures and knew the broad theory of the rotation. They were eager and enthusiastic about it and did the best they could under the circumstances to advise the farmers for whom they were responsible, but they were in an invidious position. Their technical training was in direct conflict with the principles they needed to take to the farmers who were to use the medic system. They had not been trained to consider costs of a farming system to be important and they were unsure of how to present the costs and returns to the farmers. Indeed, they had, under the past socialist approach to agriculture, not had to persuade farmers; they were required merely to direct the managers of state farms and cooperatives to carry out the technical advice emerging from the ITGC which simply repeated the requirements of the old system. In addition, their own grasp of the day to day operations of a medic/cereal rotation was shaky to say the least – even the most knowledgeable of them had not actually operated it and few had seen the total system in operation.

They faced tremendous conflict when they saw the old system still being used to prepare all crops on their stations and yet they were being asked to induce farmers to adopt another. This was epitomised when the long established spontaneous medic on the 'El Zachariah' station was ploughed up by deep ploughs to provide a traditional bare fallow in preparation for a cereal crop. This station was one of those that had been given the crucial task of advising farmers how to replace their fallow with medic pasture. The director did stop the ploughing of this particular piece of medic but on a subsequent inspection it had virtually disappeared due to overgrazing and general neglect.

Without a well-constructed demonstration and training program to back them up technicians could not effectively act as advisers. In an attempt to overcome this lack the audio-visual kits produced by FAO were made available to the technicians who were invited to use them and suggest ways in which they could be refined to suit particular Algerian conditions and then distributed to Algerian stations and research centres.

It was also recommended that technicians be made familiar with simple farm budgets to enable them, with the farmer, to illustrate and understand the savings in costs and the increases in productivity possible with the medic system. This would provide the technicians with some farm management expertise and give them more confidence in their own abilities (Chatterton & Chatterton, 1988).

Farmers and technicians in the Tiaret region after 1988

The privatisation of the farms without an accompanying change to a more environmentally stable farming system led, however, to a grave deterioration in the condition of the cereal zone as farmers rejected (through lack of resources) the old system but were given insufficient support to change to one more within their scope.

Many of the farms involved in the ITGC program were in the Tiaret region, always considered to be one of the best cereal growing zones of Algeria. We had made a number of visits to Tiaret and surrounding districts since 1979 and had seen medic growing abundantly in trial plots and on farms in the establishment phase.

The ITGC medic program had hoped to finance most of the development phase by involving various funding agencies and international research centres like IFAD, ICARDA and ACSAD. Algerian funds had been earmarked for loans to farmers to purchase plant and machinery and livestock, but problems with administration left most of them waiting for the much needed cash and they had to sell their flocks and revert to the simplest form of exploitation (wheat after wheat and minimal cultivation) in order to survive. Many were in dire poverty and often only reaped enough grain for the household requirements.

By 1990 the farming land surrounding Tiaret was in a critical state of erosion and more permanent desertification was evident in the form of incipient sand dunes. Following a deficit of about one third of the average rainfall in the year prior to the tour of inspection few crops had been reaped and when we saw it almost all the land was in fallow and ploughed for sowing in the late autumn. When the rain did come it simply ran off causing deep gullies across the ploughed land, forming lakes at the bottom of declivities and leaving the dry sown seed relatively untouched. This was not on an odd farm, but happened everywhere.

The program being offered to the farmers to overcome their problems continued to reflect the elaborate and costly cultivation program which the station used to produce its own cereals: deep ploughing, repeated cultivations, nitrogen fertiliser, and machine seeding. Most farmers rejected it and simply used a tandem disc (cover cropper) to go over the ground – once to open it up, and the second time to cover the hand-broadcast cereal seed. Few were using fertilisers and most planted cereal after cereal and ploughed a bare fallow when there was a drought. Farmers said they simply did not have resources to do any different. Few sheep were left on farms because the only feed available was purchased hay and grain and farmers did not have the cash to buy either.

The 'El Zachariah' station was keen to have farmers to adopt a rotation of wheat with a forage crop or grain legume in place of fallow in order to build up livestock production again, yet all cereal crops on the station were planted after bare fallow and forage crops were planted separately, not as part of a rotation. Not one farmer had followed the advice.

On part of the station a miniature medic/cereal rotation (under the direction of Dr Kamba, on secondment from ACSAD) was in operation using shallow cultivation and superphosphate. For four years it had produced well, had provided grazing for a small flock of sheep and the medic was regenerating. These results were not incorporated into the farm advisory program.

The technicians and extension advisers were frustrated and angry and well aware that the complex and costly system they were required to present as desirable to the farmer was unacceptable and logically unsound, yet they could not use the medic system as an alternative. The ITGC administration in Algiers had proposed and was supporting the plan to convert farms to a medic/cereal system but the direction on site was committed to the old system and could not accept that change was necessary.

Resistance by technical experts to new technology

Conservative technical experts (not all local, many were foreigners on secondment from European institutions and agencies) with local power were a barrier to those more progressive administrators and technicians who were open to new ideas and who saw that the old techniques had not only brought about the erosion and loss of fertility that were undermining farm production in the district, but were making it worse.

The demonstration on the station of a medic/cereal rotation illustrated that it was within the resources of most farmers – it was not perfect but it was better than the system they were using – yet it remained marginal because no elements of it were included in the advice the technicians were taking out to the farmers.

The sticking point proved to be the change from deep ploughing to shallow cultivation, both to lower farmers' costs and to enable them to grow medic. The technical direction insisted that deep ploughing must continue because it enabled the roots of the cereal to find moisture and grow strong and they refused to allow the technicians to advise shallow cultivation. The irony of it was that the farmers were nearer the operation of shallow cultivation than they had ever been because their poverty led them to abandon the deep plough for the sake of cheapness and to cultivate the soil with two passages over the ground with discs. The farming of desperation was widespread in the district. The technicians working face to face with the farmers could see this, but their immediate superiors were too timid to take the first step towards improvement.

The technicians were left a little like the South Australian Department of Agriculture advisers in the late 1930s who were telling the farmer to grow more pasture, while at the same time advising him to use bare fallow before cereals.

The nearby rangeland

Separate from the cereal/livestock project, on the rangeland of which the centre was Ksar Chellala with an average rainfall of 200–250 mm, some thousand hectares had been sown to medic to provide feed for a livestock production centre being developed by a private corporation. The medic was sown by an Algerian technician who had supervised the establishment of medic pasture at Domaine Chouhada during the FAO/MARA project of the 1970s. Several cultivars including Paraggio (*M. truncatula*) and Paraponto (*M. rugosa*) were planted in separate blocks using a rate of 10–12 kg/ha following two shallow (5 cm) cultivations of soil and simultaneous dressings of 100 kg/ha of triple superphosphate. The medic had not been sown until December due to pressure of work on the technician. The pasture had germinated well but was still (in January) a little too short to be grazed. The technician told us that he was thinking of oversowing the medic with barley because he feared that if the medic failed the production unit would not be pleased and that, at least, the barley would provide some feed (Chatterton & Chatterton, 1990).

Audio-visual extension material in country centres

The lack of adequate information available to technicians in Algeria about the management of the medic/cereal rotation has proved to be one of the major barriers to its adoption by farmers. If experienced medic farmers cannot be employed in sufficient numbers to explain the system to technicians and to help farmers adopt medic, then another means must be found. The use of audio-visual material is the next best. It is rarely appreciated within the communication units of the international agencies just how difficult it is to use even simple equipment in centres in less developed countries. It was difficult to convince sophisticated production units that videos and pulsed film strips are not useful.

We always travelled with film strips, photographic slides and books illustrating aspects of the medic/cereal rotation, with texts in the language of the country concerned, and had to use them in places as disparate as sheds where a group of farmers and a couple of technicians gathered, or at a farmers' meeting in the meeting house in a country centre, and also in the training centres on the stations.

The means of showing slides or film strips (the projector, power supply and so on) were often lacking. While we were able to carry our own projection equipment to the site, the technician responsible for extension advice rarely has access to projection equipment and certainly does not have any available when he sits in the field with the farmer. Our experience is that in the office of the Minister in the capital city, a screen and projector for film strip or slides can usually be found. Within the universities and colleges, screens and projectors are also available, but pulsed tapes (where a 'voice-over' commentary is synchronised with the run of the slide) usually come to grief because of difficulties with the projector or the accidental insertion of an upside-down slide, because once the pulsing gets out of phase it is difficult and distracting to correct this. The lack of synchronisation is not rapidly apparent and technicians tend to lose interest once the presentation ceases to make sense. At the stations adjacent to country towns and often at the local high schools in all countries in the region we rarely found a screen (most showings were made either on the wall or on a white sheet), the electricity supply for the projector (which we supplied) was often just bare wires twisted together and rarely were there blinds to cover the windows. FAO had made many film fixe projectors available in the region but we did not find one on our many extensive tours and as FAO does not produce slides (too costly), the lack of equipment means that film strips too often remained in cupboards unused. We were unable to persuade FAO to produce the medic information kits in slide form, but they do provide each film strip with a small booklet that contains both the script and a photograph (in black and white) of each frame or slide. These booklets were printed in English, Arabic and French, and although the translation in several places was not always accurate (e.g. burseem for sub-clover) they did provide a visual lesson on their own and, if used in conjunction with a projection of the material, enabled each technician to follow the story him/herself. This format can be interrupted and a particular point discussed without losing track of the total story. Individual centres were encouraged to make their own slides and build up their own kits using the booklet as a guide. The booklets are cheap to buy and can be taken by the technician to groups of farmers. The technician refers to them when explaining the operations under consideration that day, and the farmers can take them home and ponder the message in private or among themselves. All our slides were photographs of small farms in the region with the farmer demonstrating medic with his own flock and his own farm equipment. This helped farmers and technicians relate to the illustrations in a way that kits filmed on Australian farms or on carefully controlled research centre sites do not. Most farmers, in our experience, have some literacy in Arabic, some in French, and all are quick to understand pictures if they are presented in a way that tells a story clearly.

FAO did distribute some kits to capital city offices, but the distribution chain was not effective. As far as we are aware the ICARDA WANA project did not use these kits to introduce and inform farmers and technicians about the operation and management of medic pasture in the program they carried out in the Tiaret region in the 1990s.

Conclusion

Although the Algerian Ministry has maintained a consistent enthusiasm about medic for two decades and had provided a great deal of support for its own and agency projects during this time, the twin resources of scarifiers and expert demonstration have not been obtained in sufficient quantities to make the system generally accepted by farmers.

The FAO project left behind strong recommendations that scarifiers be made available to farmers because of the cost and damaging effects of continued deep ploughing. The research stations at El Khroub and Tiaret carried out comparisons between shallow and deep cultivation and conventional and combine seeders and the results supported the findings of the FAO project. Several scarifiers were obtained and used to good effect on one farm and two research stations. A large number of scarifier/combine seeders were imported by the Algerian Government but lack of training in the operation of these prevented their use.

Technical exchanges between Australia and Algeria have taken place at the institutional level, but this has not taken the knowledge of the system beyond the general theory of its operation.

Algerian technicians who are keen to assist farmers adopt medic must battle against the strong hold within their educational institutions of the French legacy of a dryland farming system evolved in Europe and the conservatism of some of the middle range technical experts responsible for directing local extension programs.

The experience in Algeria with the provision of audio-visual material to try to provide a substitute for expert farmers has illustrated that although useful if the technical apparatus is available to enable it to be used, it has limitations unless it is supported by the provision of appropriate machinery and training in management by experts.

13

On the farms in Morocco

Introduction

The Moroccans first began work with medic in 1981 on a World Bank funded project in the Fes–Karria–Tissa region, but attempts to employ Australian farmers and technical experts for it were not successful. The project went ahead independently and reports indicate that the medic grows well and has substantially increased the amount of feed on offer as grazing. However, the region has a relatively high average rainfall and is fertile, hilly country and farmers are able to grow a wide range of crops. For this reason it is considered by many that these conditions are not conducive to the adoption of a simple dryland medic/cereal rotation (personal communications). There was also a small applied research project operated by the German aid agency GTZ and the Moroccan INRA to carry out small-plot trials in conjunction with commercial field tests on selected livestock farms.

In 1985, as a result of a visit to Australia by a group of Moroccan technicians to see the system in operation a large scale farm development project, 'Operation ley farming', was begun. The purpose was to improve the productivity of grazed fallow, improve soil fertility and control soil erosion, and farms where fallow was used and livestock available were selected throughout the dry cereal zone from south of Casablanca to Safi and Marrakesh. The ministry took advice from GTZ and invested in medic seed and the employment of two expert medic farmers to assist in the establishment phase.

This initiative was begun considerably later than other countries in the region and by the time it began there was a wealth of experience available indicating that the scheme would fail unless farmers had scarifiers to carry out shallow cultivation and experts to help them understand the grazing regime required.

The introductory phase

In 1985 the first 18 000 ha of medic were sown. By 1988, 47 300 ha had been sown.

The expert medic farmers were in Morocco for three months of winter in 1985/6 and two months during the winter of 1986/7. The cultivars sown were advised by GTZ in Morocco and the seed came from Australia. Between 16 October 1985 and 7 January 1986, the two Australian farmers visited seventeen sites, sometimes planting the medic themselves, sometimes instructing local farmers how to do it, and at every site providing demonstrations, explanations and advice about the techniques used and the reasons for them. Neither farmer spoke French but they were accompanied by a Moroccan technician who did and who later travelled around with us on our inspection visits. Often up to 142 farmers attended these information days as well as technicians and officials attached to regional and local Ministry offices and students from nearby colleges.

Problems encountered

(a) Seed beds

The only implements available for the preparation of the seed bed were discs and chisels that produced rough, cloddy surfaces. This meant that weed control measures were not efficient and that the medic was sown onto a badly prepared surface.

(b) Conflicting technical advice

Within the 'Operation ley farming' program farmers were being advised not to plough deep. Concurrently there was another program on offer supported by the King which called for a 'Million Hectare Wheat Programme' based on American advice to deep plough, use fallow and apply nitrogen fertiliser. No scarifiers or combine seeders were available to enable 'Operation ley farming' to provide an illustration of the effectiveness of shallow cultivation to offset this. Project funds were only used to provide medic seed and some technical advice.

Australian farmers were advising Moroccan farmers and technicians that nitrogen fertiliser was unnecessary in a medic rotation – and counterproductive if applied to the medic itself as it inhibited nodulation. Yet farmers were also being advised by other technical experts that nitrogen fertiliser should be used even when sowing the medic. The experts were either ignorant or uninterested in the savings the farmers could make because the medic provided the nitrogen for his cereal crop. The Australian expert farmers recommended that 'groups responsible' for recommending nitrogen application at sowing be educated so that they became aware of its detrimental effect on pasture and its beneficial effect on the farm budget. A further difficulty presented itself: supplies of straight superphosphate were difficult to find and farmers were often forced to use a mixture that included both superphosphate and nitrogen (Rodda, 1986, sections 5 and 6).

(c) Technicians and resources

The Australian farmers noted that while the broad theory of a 'ley system' using medic pastures was well known among the technicians, there were many aspects of it that had not been adequately explained or demonstrated and nothing had been done for the farmers. The resources available to technicians responsible for introducing this type of farming change were inadequate. There was no comprehensive education program available to technicians and very little extension material produced about the use of medic pastures in Morocco.

The formidable task facing the Moroccan technician is apparent in, for example, the Ministry's Centre de Traveaux at Ben Ahmed where in the surrounding district there are 27 000 farmers and only 22 technicians. This district has an annual rainfall varying from 400 mm to 250 mm, considered quite good for cereal production by Australian standards. It contains 209 500 ha of which 177 000 ha are arable. Cereals are sown on 123 500 ha, legumes on 24 500 ha, forage on 3200 ha. Each year about 20 700 ha are in fallow. About 1500 ha of irrigated vegetables are also grown. Thus about 70% of the land is being continually sown to cereals. Half the farms are less than 10 ha in size and have no tractors or mechanised implements, and those that have implements have ploughs and other implements suitable only for deep ploughing. Technicians, in addition to many other tasks and responsibilities, were being asked to help the farmers convert that 20 700 ha of fallow to medic pasture yet they had no regular access to expertise to assist them with the management of medic pastures and at that time no extension material whatever (Masters, 1986, p. 2).

Competition from maize in Morocco

In addition to trying to carry out conflicting policies the technicians also had to try to overcome the common use of maize as animal feed. This crop had been introduced in the late 1960s even in areas where annual rainfall often falls below 250 mm. After the cereals are harvested, maize is planted on a winter fallow. The hybridised maize grows weakly but usually produces some corn. This is hand-harvested and then fed to animals during periods of shortage of other feed. The cost of planting, harvesting and feeding out is great compared to the cost of medic pasture regenerated and grazed both in green and pod form. However, the growing of maize has now become 'traditional' and no one is prepared to tamper with it. It seems that such uneconomic crops are accepted because they do not require the disturbance of the normal farm operations. One can deep plough the land before growing maize, it can be cut by hand, and the animals like it. The cost is not counted. Such are the irrationalities that exist in farming at times.

This does not mean to say that all medic programs should be abandoned because they are more difficult to introduce. Not all farmers grow maize, and many who do would be quick to stop once neighbours demonstrate that the same objective (feed for animals) can be achieved in a better way. The problem for the technicians is to find the resources with which to demonstrate to the farmers that there is a better way.

Innovation and adaptation

The absence of scarifiers and combine seeders posed problems for the South Australian farmers when preparing seed beds and sowing medic pastures during their demonstrations, yet these are the same problems faced by the technicians responsible for showing farmers how to establish and use a medic farming system and by the Moroccan farmers who wish to establish and use the system.

The Australian experts were forced to use whatever was to hand to carry out their contracts – cultipackers, discs, chisel ploughs, camel thorn bushes dragged behind tractors or mules when harrows were not available – and the disc cultivator called the 'cover cropper' which is the commonly used implement on Moroccan farms and which the Australian farmers adjusted in the hope that the result would not be too harmful to the regeneration of the medic.

They sowed with light imported seeders and by hand, and put out fertiliser by machine and by hand. In one case, when harrows were unavailable to cover the seed after planting one Australian farmer helped his group of Moroccan farmers assemble chains on a bar and that was reasonably effective.

With their experience the Australian farmers had the advantage of knowing the result they were after and took innovative action to try to achieve it. None of this initiative could alter the fact that the Moroccan farmers, without appropriate equipment and tied to time constraints, would simply have to continue to prepare rough, cloddy, weedy seed beds that were the major on-farm problem that the Australians encountered. In any case, the Australians only had to prepare a seed bed for the establishment of the medic pasture; the real problem would be after the pasture had been grazed and the land had to be cultivated for the succeeding cereal crop. This was when shallow cultivation was essential if the medic seed bank was to be safeguarded and good weed control obtained. No amount of 'adapting' an implement designed for deep ploughing or discing would perform these two operations efficiently. The proper scarifier is essential.

Seed production

During their term in Morocco the two Australian farmers went over to inspect the 500 ha planted to annual medic on the government farm for seed production at Jedida. It was their opinion that little would be harvested. The medic was sown 'without supervision' with the result that soil preparation was poor and there were too many weeds and self-sown cereals to allow the medic plants to thrive. In spite of this they considered that the medic pasture itself was not too bad, but pointed out that 'there is a large difference between a seed crop and a pasture' (Rodda, 1986, section 7). This was a common problem in the region where seed production was being attempted. The preparation of a level seed bed and the rigorous elimination of weeds is a necessity for the production of economically viable levels of seed production. Without the properly designed cultivation equipment this is difficult to achieve.

Farmers and technicians attempt the rotation

In June 1986, in conjunction with technicians of the Production Végétale section of the Moroccan Ministry of Agriculture, we visited the 'Operation ley farming' sites. The rainfall had been abnormally high throughout the country and even after grazing throughout spring and early summer there was a good supply of pods available for summer grazing and seed bank purposes. The cultivar Robinson (*M. scutellata*) had proven to be particularly productive although experience with it in Australia was that it was not suitable to areas where prolonged drought was frequent.

Grazing programs

Some farmers who had put their sheep to graze the green pasture through from January to the end of May were now leaving the flock to graze the dry medic straw. In contrast to past years, they had no need to buy concentrates and straw. During June others were taking their sheep off the medic and putting them to graze on the cereal stubble that was now available. The medic pods remaining on site were abundant, and many farmers interviewed said that they considered this an advantage because it would provide a good supply of seed for the next pasture. Some had taken the precaution of lightly cultivating the soil so that the pods were buried just below the surface of the soil and thus protected from unauthorised grazing by casually invading livestock. Only one site (near Safi) appeared to have been overgrazed and that was quite bare of pods. The farmer said he had grazed it heavily right through the flowering stage in the spring. Most farmers appeared, on the other hand, to have erred on the side of caution and had much unused grazing material on hand. The technicians were unsure about what to advise as a regime for summer grazing as it was outside their experience.

Preparations for the cereal crop

At the time of the review many farmers were beginning the long cultivation program for the cereal crop which would take them into the next phase of the rotation. They began by discing the ground on which the medic pods remained. Because this buried these pods, it deprived them of an extra period of grazing for their sheep after the cereal stubble had finished. The shorter program possible with the scarifier and combine seeder allows grazing to continue right up to early winter. Most had been asked by their technicians to be careful not to plough too deep, but two out of every three sites visited had a rough, cloddy seed bed that would inhibit regeneration in the pasture phase after the cereals. The technicians knew that deep ploughing was not good for medic, but could do no more than advise that farmers keep their discs 'adjusted'. Most farmers did adjust their cover croppers as asked, but it proved not very successful as in the closed position the discs could not penetrate the soil properly and thus weeds were not cut and killed. Several farmers had managed a fairly presentable seed bed, but it was obvious that unless farmers were given access to scarifiers with adequate points, they were going to be unable to supply the first requirements of the rotation – a shallow and well-cultivated seed bed.

Very few were prepared to allow the pasture to regenerate without the intervention of a cereal crop. Technicians, when asked what advice they were giving farmers, referred vaguely to the 'national balance' and the need for a 'national effort to produce cereals'. Some said that they believed that it was essential to plant cereals in order to ensure the regeneration of the medic.

Education and extension

Generally, the technicians were very enthusiastic about the rationale of a medic farming system and they were quick to understand the reasons for the answers given by us to farmers concerning specific problems. Often when a question was repeated on another farm, a technician from our group would take over the advisory role and perform it well.

Both in Rabat within the institutions and out in the regional centres the demand was for literature, extension material and demonstrations. It seemed from our interviews with the farmers and technicians that farmers were keen to use medic pastures and those who had benefited from the program had put a lot of effort into trying to retain the pasture and then exploit it, but it was also obvious that no one – either farmer or technician – was clear about how to manage the medic so that it could be exploited profitably. Never having seen it done, they had no idea of what they should expect.

In Rabat, ministry officials confirmed that many were confident that medic had the potential for successfully increasing yields on dryland farms. The question they kept putting was how they could evolve a system by which their farmers could exploit the medic and how they could acquire the expertise needed to show them what was possible. They had hoped for some assistance from the Australian Department of Trade and its associated sources. This proposal foundered due to the difficulty of funding it by means of a 'barter' of resources (in this case, superphosphate from Morocco in exchange for seeds and machinery from Australia). The international agencies, it seemed, were not keen to employ expert medic farmers to work in Morocco, machinery agencies were not keen to import scarifiers or combine seeders, and the Moroccan ministry did not have the funds to do it themselves (Chatterton & Chatterton, 1986).

Medic after 'Operation ley farming'

'Operation ley farming' was abandoned after a review carried out jointly by GTZ and the Moroccan ministry concluded that medic was not particularly suited to the farms of Morocco and that, while large farms with sheep could be improved by adopting medic pastures, the emphasis on research and development should return to grain legumes and forage crops (Amine & Jaritz, 1989).

This did not prevent a further project (funded by an international agency) from attempting to introduce medic onto small farms. Inspections of the project's farms and interviews with farmers and technicians were carried out in 1988–9 on the Abda Plains near Safi. Some of these farms had been part of 'Operation ley farming' and this provided a continuity in our study of the results of that particular program.

Four zones were reviewed – Jemma Shaim/Abda Plains, Chemaia, Sebt Gzoula, and Sidi Chekar. The rainfall varied from 400 mm on the coastal fringe to a rapidly declining average of 200 mm as one moved inland and southwards. The coastal city of Safi was the base for the regional headquarters of the Ministry of Agriculture responsible for the zones.

This review was specifically focused on the interface between farmers, technicians and extension agents so that an assessment could be made as to whether adequate resources were available to extension agents in the light of the demand by farmers for information and advice.

Small farms

The majority of the farms seen were quite small – from 5 ha to 15 or 20 ha and only one large farm of some thousands of hectares was visited for inspection and interview purposes. On the small farms each farmer had from six to fifteen sheep and usually two or three cows. Farms of around 20 ha had flocks of 40 or more sheep and a few cows. The family grew its own grain and some was sold. Haystacks were made from straw and livestock were fed from this store together with some silage, some grazing of the fields, and top-up purchases of concentrates. Some maize was grown but usually by the farmers with more than 5 ha of land. Many farmers cultivated the soil with a wooden, steel-tipped plough drawn by a mule. Tractors were common on farms of 20 ha and more and the implements used for cultivation were the mouldboard plough, the tandem disc (cover cropper), sometimes harrows, and usually a cultipacker or roller. Sowing and fertiliser application were by hand. Animal manure that had been stored and later mixed with poor quality straw was the fertiliser used. Soils on the whole were adequately fertilised and there were none of the nitrogen-poor crops that one saw so commonly elsewhere. The large farm (owned by an absentee landlord) was being badly managed and there was no use of farmyard manure.

Because of the widespread use of animal manure as fertiliser the ability of the medic pasture to replace expensive nitrogen fertiliser is therefore not as potent an incentive for small farmers in this part of Morocco as it is in Algeria, Jordan, and other countries of the region. There is also a high concentration of superphosphate in the soil in some parts of Morocco, such as Karigbah where it is mined, so that the application of superphosphate is not always necessary in order to stimulate cereals and pasture. Some medic growing on parcours on several sites in the Abda Plains district appeared to be lacking in superphosphate as its growth was stunted in spite of more than adequate germination and recorded regeneration from the original 1985/6 plantings.

Surviving the rotation

Farms on which medic had been planted five or six year ago were again inspected and regeneration (albeit declining each year) had occurred in spite of ploughing with the tandem disc in two of those years. It appears that, as in Tunisia, a strong establishment of medic will survive about three cycles of a rotation when the tandem disc is used before becoming too sparse to sustain adequate grazing.

The diminution in the number of medic plants over the period is, however, exacerbated by erroneous advice being given by the technicians that the soil from which the regenerated medic will emerge must be ploughed before the rains come to ensure the germination of the medic seed.

Several farmers were using the tandem disc in the recommended closed position to cultivate the soil for their cereal crops and on these farms the medic pastures were regenerating adequately, but on one or two the problem with weeds was becoming serious. Advice to close the tandem disc had been given in the hope that this would prevent the inversion of the soil, but, while it was relatively successful as far as the regeneration of the medic was concerned, it did not provide good weed control and the subsequent cereal crop suffered. Even if the grazing of the medic is continuous and carefully managed in the pasture phase, poor weed control due to inefficient cultivation usually forces the farmer to return to fallowing.

Technicians tended to blame the medic for the weeds in crops. Farmers, on the other hand, are well aware that weeds are endemic in cereal crops in the region. Because they have poor access to spray plants and herbicides, very little use is made of herbicides on small and medium private farms. It can be argued that shallow cultivation and careful grazing provide a more environmentally acceptable way of controlling weeds than the introduction of programs to induce farmers to invest in and use chemical herbicides.

Grazing programs

Farmers with both small and medium-sized farms were very enthusiastic about the benefits that came from grazing the medic, the major ones being a saving in purchased feed and better growth of animals. There was little summer grazing, but one farmer said that he had brought in large flocks of sheep in summer to eat the abundant pods that had remained after the winter/spring grazing. When asked why he did not leave the pods to provide seed to regenerate in the following year, he said that the summer feed was worth too much to him and that, in any case, he could get more medic seed from the government who were subsidising it heavily in an attempt to induce farmers to use it.

Some farmers had developed a type of 'shifting agriculture' where they used the medic until successive deep ploughing (about three years) reduced the plant cover. They then planted another nearby field with medic pods and seed that they gathered themselves and grazed that until cultivation again weakened it, then returned to the original field and replanted it with medic.

Many farmers were planting feed barley with medic seed and grazing both and making hay, sometimes sowing more medic seed as the component of medic plants decreased. In the case of this continuous barley/medic pasture the cultivation of the ground in preparation for the sowing of the barley was quite rough and ready – usually only a passage of the tandem disc (cover cropper) to open up the soil. On one farm where grazing was well managed it had been suggested that a herbicide be applied during the cereal phase to control the weeds a little better, but such a program was beyond the farmer and he continued to rely on grazing of the previous pasture.

Most farmers grazed the medic as a forage crop, but there were a number of farmers who had grasped the benefit of grazing it to keep the plant low and spreading and they reported a continuous source of fodder during winter and spring together with a good production of pods for the summer. Several had their sheep grazing continuously until the cereal stubble was available in the summer and they were very enthusiastic about it.

There was a belief that cattle were harmful to medic pasture because their large hooves tore up the plants, and advice that dairy cattle, beef cattle and sheep all grazed medic and sub-clover pastures for most of the year in Australia was met with by scepticism not only among farmers but also technicians.

Rangeland (Sidi Chikar)

On the rangeland south and west of Marrakesh, a rangeland regeneration was being undertaken in a zone where annual rainfall is below 200 mm but where a large amount of sheep production takes place. The conventional practice is to graze livestock on opportunistic plantings of barley, to gather the remaining low quality straw for stacking, and to feed this together with grain and concentrates for a large part of the year. The opportunistic ploughing and cropping of this flat rangeland has led to exhaustion of the soil and erosion. There is practically no natural pasture left.

In an attempt to provide some grazing, large areas of atriplex were planted – behind fences and at great cost. After five years, only one large piece had been grazed in cooperation with surrounding families of livestock owners. The atriplex had not stood up to the grazing well and stock had been withdrawn after a short period. None the less, the planting continued using nursery-raised plants and water brought by tanker. Some spineless cactus had been planted in an attempt to reduce the problems incurred when sheep were grazed on atriplex alone and ingested too much salt. This had failed as the cactus did not grow strongly and required too much water for its continued existence. Some small medic plants were seen on one site but these were too few to provide a basis for the broadcasting of superphosphate that was recommended in the case of parcours in the higher rainfall region of Sebt Gzoula. The exhaustion of the soil throughout the entire region will not be reversed by atriplex plantings sheltered behind high fences.

The widespread sowing of medic pasture (such as was undertaken in Libya and Northern Iraq, or a program similar to that undertaken in Ma'in in Jordan) could provide an alternative. However, due to the exhaustion of the soil and the absence of almost any vegetation at all, a program of testing to ascertain suitable cultivars and almost certainly inoculation measures in the first instance, will be necessary before any attempt can be made to sow on a broad scale.

Certainly, the replacement of opportunistic cereal cropping with its attendant cultivation of the soil by regenerating medic pastures would be environmentally sounder and would make more feed available over a longer duration. Hay made from medic straw would be of better quality than the present stacks of low quality cereal straw that are the only drought reserve available.

Despite the logic of this approach, there is at present in Morocco among technocrats a widespread belief that medic will not thrive on land with rainfall below 350 mm. Experience of success with medic in rangeland in Libya, Jordan and Iraq remains unknown. Dormancy of the medic seed is understood to be of value in allowing a cereal crop to intervene in the pasture cycle but the value of dormancy as a drought-evading characteristic together with the medic's ability to germinate quickly and grow abundantly when rain falls, is not widely appreciated. The improved nutrition that would be available for livestock currently fed on poor quality straw and minimum quantities of costly grain does not rate consideration.

Thus the technicians responsible for operating rangeland improvement programs are advised by technical experts to go on planting atriplex and spineless cactus, and occasionally other fodder shrubs such as acacia. They are miserably aware of the practical and ecological limitations of this advice and the fact that it has so far failed to appreciably improve the amount of fodder available for livestock in the rangeland. These technicians have not had access to reports of the experience or the results achieved in other countries in the region where medic pastures have proved their ability to produce good feed on rangeland, and they have no access to funds that can provide medic seed, disc pitting machines, and inoculum and phosphate fertiliser that would enable them to try such a program in Morocco.

The demoralisation of technicians

The most striking impression was the demoralisation of the technicians and extension agents who felt abandoned by their superiors and their peers as they struggled to develop their own knowledge of the behaviour and management requirements of medic pastures with no written or visual references available to help them.

They had been instructed to introduce 'ley farming' – the ideal of one year medic, one year cereals – and not unexpectedly they had not achieved this. They were well aware, as were many of their peers and the farmers, of the broad principles of the medic/cereal rotation, and they were also only too well aware that without proper implements they could not expect farmers to take their advice not to plough deeply when preparing ground for cereals. There was only one scarifier in the region and that was on the research centre farm where it was scarcely used. Only one or two of the technicians had seen a medic/cereal rotation in operation (and only a part at that, briefly in Australia) and, while they had a keen appreciation of the logic of its applicability to the farming zones in which they were working, they were baffled as to how they could encourage farmers to use it when they had so little technical and management expertise themselves.

The extension agents were few in number and had only small motor cycles on which to visit farmers. Summers are very hot and in winter there is thick, sticky mud on all farm tracks. They were willing to hold seminars and workshops, even meetings of groups of farmers in the fields, but what could they tell them? What could they show them?

Farmers and technicians in alliance

Several technicians were dynamic enough to ally themselves with progressive farmers who had not only established medic pastures on their land in 1985 but were discovering ways and means of exploiting it in a profitable manner.

Surprisingly there was a large amount of medic being grown on farms in all the districts of the Safi region. The official version was that little, if any, existed. The technicians (in support of their claim for more resources for their own role in creating interest in medic pastures) wanted to show fields where medic had entirely disappeared but were unable to do so. On every occasion the farmers explained that they had lost their original plantings due to deep ploughing during the cereal phase but they would then go to another field where medic was growing abundantly. Farmers were re-sowing the medic, with collected pods from both sown Australian varieties and the local cultivars that grew spontaneously because they believed it to be a valuable pasture plant and they were using it as winter and spring grazing, sometimes for hay, and sometimes in conjunction with barley. The statistics supplied through various agencies showed a decrease in the demand for medic seed from the government agency but these figures were clearly not reflecting the real situation.

In Rabat it was suggested that the initiative of those farmers who were collecting and using their own medic seed might be usefully incorporated in a project to encourage other farmers to do the same, but this was not greeted with enthusiasm. The response was that it may interfere with the SONACOS monopoly of seed marketing.

Given the ubiquitousness of medic observed in this winter (December) review of several hundreds of farms, it seemed likely that medic in this region of Morocco is beginning to occupy the place it once had in Australia – an 'adventitious' plant that farmers realise can provide quite a lot of good, cheap feed, but its presence remains formally unrecognised because it has not been introduced onto the farms in the conventional way that scientists and technical experts prefer. Unfortunately the persistence of deep ploughing must continue to have a negative effect on this adventitious pasture.

The argument for and against re-sowing medic regularly

Farmers under the 'Operation ley farming' program were sold medic seed by SONACOS for 9.85 MD/kg (US$2), which represented a subsidy of or about one half to one third of the world price at that time.

Due to the subsidies some farmers were allowing their sheep to consume all the dry pods and straw in the summer and were prepared to re-invest in more seed in the subsequent year. Others who lost the medic following two intervening cereal crops sown following deep ploughing were also quite happy to re-sow their pasture.

In Australia in 1986 it was seriously being considered whether re-sowing medic pastures every few years (or even after each cereal crop) may be economically sound given the desire to achieve the optimum productivity available from pure medic pastures.

The relatively cheap investment required to import medic seed into Morocco together with the high returns from sheep and the avoidance of any need to radically change the existing cultivation program suggests that re-sowing could be a sound commercial decision for some farmers in Morocco. For the small farmer in these marginal zones, however, there are other factors to take into account.

- If the land is cultivated each year then the erosion that is currently of such concern will be exacerbated.
- The seed bank will be wasted.
- If the land must be cultivated each year then the savings associated with the regenerated pasture are lost.
- If the medic is re-sown each year the profitability of the pasture is lessened because some productivity is lost because of shorter growing period of sown as against regenerated medic.

Economic comparisons between the two farming systems

Very few comparative data of costs and returns have been accumulated and none were available to technicians. This was the same problem faced by technicians in Tunisia and Algeria. Farmers in Morocco who had integrated a rudimentary medic system into their enterprise were quite confident about its savings and benefits and this is why they were continuing to make small changes to incorporate it into their systems, but the technicians and extension agents who were responsible for attracting other farmers to use it had no idea of the relative costs and returns involved. When this was discussed in seminars the technicians were keen to gather these data for their extension programs. The collection is easy and provides them with a sound base on which to formulate their role *vis-à-vis* the farmer. It allows them to adopt the role of 'expert' without exposing their lack of self-confidence about the technical realities of shallow cultivation and winter and summer grazing. Those who do have self-confidence in these aspects of medic become even more adept at convincing the farmer about the benefits of change.

Interviews undertaken to establish relative costs and returns of medic *vis-à-vis* other systems revealed that within the research and scientific centres in Morocco there is not a widespread sensitivity about the cost effectiveness of resources on farms, whether they be small or large farms. The opinion that medic was only economically viable on large farms like those in Australia with large areas and a great deal of capital had taken firm root among administrators interviewed. Experience that conflicted with this opinion elsewhere in the region (Tunisia, for example, on farms of 5 ha, and in Libya on farms of 80 ha) did not seem to be known.

The technicians' response to lack of support

Because it seemed impossible to successfully overcome the problem of poor cultivation and subsequent difficulty with regeneration, technicians generally dropped the idea of the rotation and simply delivered the medic seed, helping the farmer where possible to understand what little they themselves understood of the grazing management required. Several of them grasped the concept of continual grazing and the consequent benefit of healthy tillering and good pod production and where they were able to convey this to the farmer the benefit was clear to see (Chatterton & Chatterton, 1989).

Conclusion

The Moroccan 'Operation ley farming' was a bold attempt to leap from the trial plot to the farm, but if one compares it to a similar bold attempt in Libya in the 1970s one can immediately see that the absence of a critical farm implement and expert demonstration of its management doomed it to partial success. The Moroccan Government invested something like US$4 million in seed and approximately US$40 000 on farmer expertise. Had they invested US$2 million in seed, US$1 million in scarifiers and US$1 million in farmer expertise they well may have succeeded in their objective.

The struggles farmers made to overcome their lack of scarifiers and lack of expert advice about the management of the system indicates their desire for a way out of their present dilemma. The government appears to have been convinced that once the plant was established all else would follow. This proved to be fallacious.

The farm management chart (Fig. 11.1) illustrates how dependent farmers are on the scarifier and correct grazing management if a productive medic/cereal rotation is to be operated on the farm. Their absence illustrates how little understanding there is within powerful institutes and agencies about the differences between a medic farming system and the conventional system.

As in nearly every other country in the region, the technicians involved in 'Operation ley farming' were given a theoretical message and then left to rely on their own energy and common sense to turn this into advice to farmers who wished to use medic. The education they receive at agricultural college teaches them that deep ploughing and bare fallowing are essential and leaves them without any information about possible alternatives. A seminar or two and a look at an audio-visual kit is not sufficient to make them expert enough to advise farmers how to manage a medic farming system.

As a result of 'Operation ley farming' the residual interest at the farm level and the ingenuity of some farmers and a few technicians may result in some useful exploitation of medic pastures taking place in Morocco. Unfortunately the review by Amine & Jaritz, with its pessimistic assessment of the program, is more likely to be taken seriously by funding agencies than the small successes of low status technicians and inarticulate, out-of-sight farmers.

While what is happening on the farms continues to be ignored, little progress will be made to overcome the erosion and the deficits in cereals and livestock feed that cause alarm.

14
The future for medic

Introduction

Between 1974 and 1983 Australians grew medic pastures either alone or in a continuous rotation with cereals in conditions as diverse as the arid rangeland of Libya and the cold cereal zone of Iraq. Some knowledge of the value of medic pastures as a replacement for bare fallow now exists in institutes and agencies outside of Australia but the farming operations that turn it into a farming system are not yet sufficiently understood and this has prevented its widespread adoption onto farms.

Local technicians responsible for persuading farmers to incorporate medic pastures into their farming systems have found it impossible to achieve success. They remain baffled about what they are to say to farmers to persuade them to stop doing what they now do, and adopt the operations and management used in a medic system. The attraction of the low cost of the medic system to farmers is not appreciated within the various agencies whose active participation in agricultural development enables them to either block or facilitate new programs and access to new resources. Yet the system evolved and became widely used in Australia because cash-poor farmers coping with poor soil and a semi-arid climate could only continue farming if they discovered a simple, cheap and efficient farming system. The medic system fulfils all these criteria. It flourishes best in dry conditions, it costs little and it allows flexibility in response to seasonal conditions. These are the criteria that farmers relate to and these are the criteria that expert farmers concentrate on when working side by side with their colleagues in other countries. These criteria also dictate the implements needed, the grazing regime proposed and the overall management of the system but they are not criteria usually taken into account by technical experts. The recommendations for expert medic farmers to be employed to show how the system is established and exploited have rarely been acted upon.

Institutions and agencies who do not put a high importance on the costs of farming operations and the savings that can be made from alternatives remain heavily in favour of grain legumes and forage crops as a replacement for fallow. Farmers have voted against these because they are too costly and time consuming. It is unfortunate that the institutions appear to be voting against medic pasture as a replacement.

Why the negative attitude?

When the early champions of medic pastures presented their case, the strength of their argument was often diminished because after presenting strong empirical evidence of the suitability of a medic farming system to the region, they went on to hedge it around with a mass of conflicting opinions about how to get it on to local farms. These often contradictory opinions were seized upon by those who believed that the system was too complex to proceed with. These opinions often came out of project experience where medic had been used as a forage crop and not as a pasture within a farming system. The fundamental difference in the management of a pasture and a forage crop were not always recognised, let alone demonstrated.

Almost every project demonstrated that a scarifier is necessary to provide effective weed control and prepare a good seed bed. Every project report criticised the rough, cloddy seed bed and the effect on erosion of cultivation using discs or ploughs. Little effort was made to bring about a change in the type of machinery being used on farms, and projects continue to list inappropriate machinery on their acquisition list and to bemoan the result in their reports.

The conventional path to introducing new technology into national institutes was tried, that of sending students to foreign universities to undertake masters degrees and gain Ph.D.s in scientific agriculture. This has not been a useful tool as far as medic farming is concerned. The majority of the students who have returned with an Australian diploma or degree have been lost in the maze of their respective ministry's service and seldom have they been employed in a medic program. In addition many of them say that they are isolated if they do continue to study medics and most of them soon take the conventional career path, becoming technical experts or trainee scientists in aspects of the current system in order to progress up the ladder of promotion. The few who stick to medics do so within the research centres and concentrate on identification and selection programs for cultivars of medic. The few technicians who are sent to advise farmers about medics usually have a diploma from an agricultural high school that teaches only that deep ploughing and bare fallow are essential and that shed feeding is the way to go with livestock. Those who have undertaken rangeland studies overseas are caught in the atriplex/spineless cactus, stock removal syndrome. Many agricultural high school staff and principals are eager to introduce aspects of medic farming into their courses and offer cooperation if their ministries will make funds available to carry out practical and theoretical instruction for their students. The matter is enthusiastically greeted in the capital but funding somehow never appears. The degree of penetration achieved by the Australian farmers in Libya and Morocco into the consciousness of the farmers and technicians has been much greater than any program that has relied on facilitating overseas degrees for a few technicians.

While projects continue to avoid dealing with these matters, attempts to transfer the system will remain in limbo.

A Kuhnian revolution?

Perhaps the time has come for scientific experts involved in dryland farming to undergo a Kuhnian revolution? Agricultural science is facing unprecedented criticism of many of its incursions into developing countries with climates unfamiliar to those who provide advice. In studying the introduction of the medic system from Australia to other dryland zones one is struck by the narrowness and conservatism of the international agencies and the networks they support to carry out technical assistance. Although there is much to-ing and fro-ing of individuals between countries in the name of technical cooperation, there is little attempt to make connections between identified problems and logical solutions. The mantras that haunt agricultural colleges and universities, the jargon that emanates from farming systems experts in love with modelling, are too easily accepted and passed from one generation to another with an absence of intellectual rigour that is reflected in the way in which data and experience are simply ignored or rejected if they do not reflect the clichés trotted out at international conferences. This lack of intellectual rigour reinforced by institutional weight is a great barrier to effective change and, in the case of the medic system, it is particularly so. The institutions and agencies have played around with variations of a medic farming system for nearly half a century trying to understand how and why it works and yet the simplicity of its operation and the reasons for its value to dryland farmers has evaded them.

Some farmers in North Africa and the Near East who have had some exposure to the ideas and who have been recipients of supplies of seed are fumbling towards their own medic farming system. Surely it is not good enough to stand aside and let the use of medic pastures filter through the farming community forcing farmers in North Africa and the Near East to re-invent the wheel on their individual farms.

The ecological disaster that is overtaking farming and rangeland, particularly in North Africa and the Near East, is critical. The cereal farms of Tiaret in Algeria and the rangeland of Syria are vivid examples of the way in which those who work the land are dangerously at risk of losing their livelihoods altogether. This has ramifications for the growing populations within and on the edge of the cities who cannot produce their own meat and grain and who depend on farmers for their supplies. Fears of falling production always emerge when a prolonged drought deprives people of regular supplies of food at accessible prices, but if the ecological trend towards widespread desertification continues on farms and rangeland then good rains will not put the matter right.

The institutions and agencies believe that the answer to the adoption of medic on dryland farms lies in the distribution of the correct plant and this is where they are putting their greatest effort. They are wrong. Farming is a complex operation and the improvement of a plant species is only useful if the environment exists for its exploitation. Experienced farmers who have used medic pastures alone and in rotation with cereals in semi-arid conditions know what is needed for successful exploitation, why it is needed, and how to do it. They hold the key to the transformation of a useful plant into a farming system.

Medic in the United States

As an example of what happens, if this key remains unknown is the contrast in the development of the medic system in Australia and the lack of development in America. In the 1930s American farmers had at their disposal widespread knowledge of the value of medic as livestock feed and American scientists had gone some way towards dealing with the problems of seed supply and inoculation. We have the record of California Bur-clover (*M. denticulata*) which was in 1937 identified as 'one of the few Western examples of an aggressive and valuable introduced forage plant'. It was supposed that it was introduced into southern California from Spain and had become not only abundant in California but spread and was growing vigorously from 'Nova Scotia to Florida, throughout the South and Southwest, inland to Nebraska, south into Mexico and on the Pacific coast northward to Washington'. The growing habit of the species is described exactly as the medic was being described in South Australia.

Medic inhabits practically all soils such as sands, stony loams and adobes provided they are moist and not wet . . . in California the plant may occur in pure stands but more often is associated with other winter annuals . . . bur clover is one of the most valuable annual forage plants on the Pacific coast and although the green foliage has a somewhat bitter taste . . . all classes of livestock except horses and mules eat it greedily, especially when it is maturing. At this time it is very nutritious, where abundant, it serves as a finishing feed for lambs . . . because of abundant and nutritious burs this species provides a summer and fall feed superior to that of other common annuals; sheep especially, seek the fallen burs . . . chemical analyses indicate that California bur clover is very similar to alfalfa in forage value and definitely superior to the former's common annual range associates, especially in proteins and phosphorous content . . . Like other forage plants bur clover loses nutritive value . . . after it cures on the range, although the burs are not appreciably affected . . . Bur clover is invaluable as a cover crop or green manure . . . because of its rapid, dense growth and its ability as a legume to increase the available nitrogen supply to the soil.

Not only was medic used by farmers and known to scientists in the United States in 1937, but scientific experiments had been carried out to assess the digestibility of medic and its nutritive and agronomic value. It was being considered for rangeland revegetation, viz: 'This plant has considerable promise for range reseeding at the lower elevations where mild winters are the rule'. In addition, just as Hannaford in Australia was garnering his seed supplies from the impurities ejected from cereal seed, so American farmers were able to obtain seed readily because 'it is a common impurity in grains, and is also salvaged from wool wastes' (*Range Plant Handbook*, 1988, pp. 433–5). However, the regenerative characteristic of the plant and the value of manipulating this characteristic to create a cereal/livestock system was not known or, if it existed, was not recorded. Whether American farmers continue to exploit medic as a grazing source is also unknown but they certainly have not used it in rotation with cereals so the knowledge so carefully recorded by scientists in 1937 has not led there to the farming system now so common in South Australia.

This can be deduced from a comprehensive paper on dryland farm erosion in which William Lockeretz discusses the dustbowl problem in the Great Plains of the United States and suggests that no strategies have yet been discovered to deal with the recurring problem other than an ill-fated pivot irrigation scheme that caused even more erosion because of the removal of shelter belts to accommodate the sprinklers, attempts to reduce cultivation of summer fallow, and now taking some land out of production – a plan Dr Lockeretz claims is subject to sudden change when markets cry out for more wheat (Lockeretz, 1981, pp. 140–9).

Barriers to farmer employment

Why are farmer experts who hold the key to the medic farming system not being made available to other farmers with problems of declining yields and increasing soil erosion? Partly because farming expertise is not valued greatly by agencies and it has proved extremely difficult to acquire adequate demonstration and training for technicians and farmers involved in medic projects.

The tension between the technocrat and the expert farmer is stubborn. Both groups feel superior to the other. The technocrat is open about his belief in his own superiority, the farmer is more covert and only admits his feelings to other farmers. Of course the level of intelligence and innovative energy differs among farmers as it does within groups such as scientists, academics, technicians and administrators, but one can make a special case for the expertise of the South Australian farmer who carries within himself such a unique and comprehensive knowledge of the profitable exploitation of medic pastures.

The medic farming system grew out of farmer innovation. The farmers who evolved it were an aberration – they discovered it in a moment in time when they ruled their small world if only because they were living in a country in which the rest of the world had little interest and influence. They only had this power for a short time – fifty years at most – but this was enough to enable them to establish the fundamentals of the system and to provide them with enough confidence in it to withstand the assaults that technical and scientific experts mounted against it during the next fifty years. Their descendants have inherited this confidence and have added their own experience and expertise to it. They have shown that they are able to work sympathetically and constructively with farmers in North Africa and the Near East. They have reversed the advance of soil erosion, the decline in soil fertility and the degradation of marginal land on their own farms and in Libya. Attempts to use other means of transfer and other models of technology to solve these problems have failed. Without doubt they hold the key necessary to understanding how to conduct a successful medic farming system but if the agencies and institutions reject them can they be given the opportunity to share it with their fellow farmers in some other way?

The marginality of Australian farming on the world scene

Given the minor place of Australia on the world stage, and particularly in the world of agricultural technology, it has been said that if a medic farming system is to be set up in the region it will have to be re-invented in Europe. One can already see this happening in occasional reports emanating from Europe of trials assessing the hard seed component of medics, the nitrogen fixing capacity and other botanical aspects of the plant. It is highly dubious that this replication of work already done in Australian institutions and on projects and research centres within the region will catalyse a better farming system on dryland farms where soil is capping or blowing away or on rangeland where the pasture plants are fast disappearing.

These studies alone will hardly encourage the type of funding needed to support the introduction of a new farming system to the region. A powerful donor is needed.

Throughout the region one sees the influence that powerful donors exert in determining the type of farming technology on offer. For instance, the United States focuses its solutions to problems of dryland farming by trying to make deep ploughing more efficient, by demonstrating chemical fallows, and by encouraging nitrogen fertiliser, because these are preferred by the institutions and used by farmers in the United States for cereal production and they recommend grasses and shrubs to improve rangeland. The same applies to Holland, Germany, France and the United Kingdom.

National networks within development agencies also determine the type and origin of farm machinery that goes into aid programs, the nationality and status of experts who work on aid progams, and the content of the program and its objective.

Australia is the only country where the medic farming system exists but Australia has only a small network and it plays a minor and quite feeble role in the big aid world. Few Australian technocrats working in the big agencies know much more about dryland farming other than that it is supposed to be something Australia is good at. So superficial is the knowledge of it at administrative levels in Australia itself that the national aid program, while acknowledging that Australia has unique expertise in dryland farming in a Mediterranean climate, does not support programs using this expertise as they are outside the sphere of political influence Australia wishes to be involved in (Chatterton & Chatterton, 1988, pp. 171–88). Instead, the Australian aid agency (AIDAB) and its associated research institute (ACIAR) concentrate on providing technical expertise to tropical and subtropical zones in China and the Asia Pacific region – with a few token programs in East Africa and one for the islands of the Indian Ocean. The ACIAR management's geographic mandate declares that 'projects in regions such as Western Asia and North Africa are considered only in very special circumstances' (ACIAR Annual Report, 1987–8, p. 5).

Few if any Australians working within the multi-national agencies in Europe or the United States follow career paths that are directly concerned with transferring Australian dryland farming expertise to dryland farmers in other regions of the world with a Mediterranean climate.

The individual Australian state governments have blown hot and cold over dryland farming projects. The Western Australian Government lost interest after the Iraqi project ended, and while the South Australian Government continues to send representatives to North Africa and the Near East to look at possible projects, none have successfully completed the necessary negotiations. Attempts by some South Australian traders and Seedco farmers to interest the Australian Government Department of Trade in a combined effort to market dryland farming expertise came to nothing after many months of conferences and several draft proposals (Austrade Draft Strategy, 1987).

Some farmers in Australia do not support the transfer of the system to North Africa and the Near East because they fear the threat that its successful adoption may cause to their markets in the region for sheep and grain. Some farmers are not happy to see their seed production techniques being transferred to the region for the same reason.

Australians have succeeded in bringing a broad knowledge of the system to farmers in other dryland zones and have shown that medic pastures do provide a stable and productive farming system that reverses environmental degradation and increases production of livestock and cereals, but Australia does not have the international muscle to do much more.

Farming development instead of technical cooperation
If the thrust of change is tied not to technical cooperation but to farming development, then the emphasis must change from the research centre to the farm. The Jebel el Akhdar Authority project placed the emphasis on the requirements of the farm and the farmer in order to create change and this proved a successful model to emulate.

This emphasis, with the possibilities it provides for trade and influence, may well activate a more powerful country than Australia to take a direct interest in funding the necessary projects and programs. Already much of the farm machinery used only in Australia is manufactured by multi-national companies with head offices in Europe and America so design patents should not inhibit such a development. One could envisage Italy, for instance, with its well developed machinery manufacturing sector and its own Mediterranean farming zone becoming a major force in supporting practical programs for the introduction of medic/cereal systems using farmer experts and Australian designs for cultivation and seeding equipment but Italian manufacture. France, with its special involvement in North Africa may well see the benefit of adopting medic pastures as a means of increasing its influence and trade in the region. France has the capability to produce medic and sub-clover seed in commercial quantities and it already sells considerable forage crop seed in the region. France also has a large manufacturing industry and the capacity to tool up for implements suitable for shallow tillage and precision seeding. The influence of French textbooks for agricultural education has been enormous and all pervading in North Africa and it is here that the introduction of the theory and components of the use of medic and sub-clover pastures on dryland farms and the rangeland are essential. Will the French produce such textbooks or make the changes required to the present generation of texts, or must it come from elsewhere? Great Britain on its own and through the EEC has for many years had a considerable influence within institutions and development agencies in the Near East and it is not inconceivable that British industry may take up the challenge of producing scarifiers and combine seeders. These machines would be assured of finding a good market if Great Britain, through its agricultural institutions, supported the marketing campaign with the insinuation of the knowledge of a medic farming system into programs and courses. Germany has established a valuable sphere of influence through its aid agency GTZ which has projects or representation throughout the Near East and North Africa, strongly supports agricultural machinery imports and did pioneering work with sub-clovers in Northern Tunisia and some work with medic in Morocco.

If expert farmers from Australia can join with a powerful donor country determined to carry out a dynamic program to provide the implements and the know-how necessary to benefit from a sustainable medic farming system, then the declining productivity and erosion of dryland farming not only in North Africa and the Near East but in other dryland zones can rapidly be reversed.

References

Chapter 1

Adem, L. (1989). Synthèse de la recherche sur les espèces annuelles de medicago en Algérie. In *XVI Congrès International des Herbages*, Nice.

Allan, J. A. (1989). The effects of the demand for livestock products on natural resources. *Libya: State & Region – a study of regional evolution*, eds. Allan, J. A., McLachlan, K. S. & Buru, M. M. School of Oriental and African Studies, Centre of Near and Middle Eastern Studies, London University, and Al Fateh University, Tripoli, and The Society for Libyan Studies, London, pp. 119–26.

Anon. (1987). *La culture intensive du blé*. Institut technique des grandes cultures, Ministère de l'Agriculture et de la Pêche, Alger.

Boutonnet, J.-P. (1989). *La speculation ovine en Algèrie in produit de la céréaliculture*, Série Note et Document 90, Montepelier.

Carter, E. D. (1974). *The potential for increasing cereal and livestock production in Algeria*. Report for CIMMYT, Mexico and Ministry of Agriculture and Agrarian Reform, Algeria.

Carter, E. D. (1975). *The potential role of integrated cereal–livestock systems from southern Australia in increasing food production in the Near East and North Africa region*. Report for the FAO/UNDP regional project on the improvement and production of field food crops. Waite institute, Adelaide.

Chatterton, B. & Chatterton, L. (1982). The politics of pastoralism. *Australia Habitat*, **10**.

Chatterton, B. & Chatterton, L. (1986). Conserving native vegetation on private farms in South Australia. *Australia Habitat*, **14**.

Chatterton, B. & Chatterton, L. (1987–90). Field notes for supervision missions to projects at Le Kef–Siliana in Tunisia, Abda Plains in Morocco, and Tiaret in Algeria. Chatterton Papers, Italy.

Cocks, P. S. (1984). Ecological study of annual legumes in the marginal lands of Syria and Jordan. *ICARDA Annual Report*, Syria.

Donald, C. M. (1964). Phosphates in Australian agriculture. *Journal of the Australian Institute of Agriculture Science*, **30**, (2), 75–105.

FAO Yearbook (1990). *Trade*, **44**. Rome.

FAO Yearbook. (1991). *Production*, **45**. Rome.

FAO. (1992). *Agrostat data base printout, 21 December*. Rome.

Gintzburger, G. (1980). *Rangeland Regeneration in Libya*. Report to the Secretariat of Agriculture, Reclamation and Land development. Socialist People's Libyan Arab Jamahirya, and FAO, Rome.

Gintzburger, G. & Blesing, L. (1979). *Genetic conservation in Libya. Indigenous forage legumes collection in Northern Libya (Spring 1978). Distribution and ecology of* medicago *spp.* Socialist People's Libyan Arab Jamahiraya Agricultural Research Centre.

Jamil, M. & Karem, H. (1987). *A Study about Socio-Economic Farming Systems in Ma'in Pilot Area.* Ministry of Agriculture and Forestry Department, Amman, Jordan.

Perkins, A. J. (1927). Ten years progress in wheat growing. *Journal of Agriculture,* **31**, 240–53.

SAYB (1970) *South Australian Year Book, No. 5.* Commonwealth Bureau of Census and Statistics, South Australian office.

SAYB (1985) *South Australian Year Book, No. 20.* Australian Bureau of Statistics, South Australian office.

Seedco. (1988) *Seedco, Seeds of success, A twenty five year history.* South Australian Seedgrowers' Cooperative, Adelaide.

Squires, V. (1981). *Livestock Management in arid zones.* Inkata Press, Melbourne.

Swearingen, W. D. (1988). *Moroccan mirages, agrarian dreams and deceptions 1912–1986.* I. B. Tauris, London.

Trumble, H. C., Whyte, R. O. & Nilsson-Leissner, G. (1953). *Legumes in Agriculture.* FAO, Rome.

Tyers, R. & Anderson, K. (1992). Disarray in world food markets, quoted in *The Economist,* 12 December, 1992.

UNESCO–FAO (1963). *Bioclimatic Map of the Mediterranean Zone.*

WAOPA (1985) *Final report and recommendations, development of dryland agriculture, Jezira project, Northern Iraq.* Western Australian Overseas projects Authority. Government of Western Australia, Perth.

Webber, G. D., Cocks, P. S. & Jefferies, B. C. (1976). *Farming systems in South Australia* South Australian Department of Agriculture and Fisheries, Adelaide (also available in French and Chinese).

Williams, M. (1974). *The Making of the South Australian Landscape.* Academic Press, London.

YBA (1984). *Year Book Australia.* Australian Bureau of Statistics, Canberra.

Chapter 2

Bagot, C. H. & Ridley, J. Correspondence 1844–77. Letters to and from Bagot to Ridley and his daughters and other miscellaneous references to the invention including a plan of the original and a copy of the report of the dedication of the Ridley memorial printed in *The Register* of 15 March, 1915. South Australian State Archives, Adelaide.

Black, A. W. & Craig, R. B. (1978). *The Agricultural Bureau: a Sociological Study.* University of New England, Armidale.

Bull, J. W. (1878). *Early experiences of colonial life in South Australia.* The Advertiser, Adelaide.

Callaghan, A. R. & Millington, A. J. (1956). *The Wheat Industry in Australia.* Angus & Robertson, Sydney.

Capper, H. (1838). *Capper's South Australia containing the history of the rise, and progress of the colony, hints to emigrants and a variety of useful and authentic information, embellished with 3 maps showing the maritime portion of the located districts and the surveyed district of Adelaide and Encounter Bay and City of Adelaide.* Published by the author at South Australian Rooms, 5 Adam St. Adelaide.

Chambers, R. (1983). *Rural Development – putting the last first.* Longman, London, Lagos & New York.

Chatterton, B. & Chatterton, L. (1981). Drought – the problems of policy implementation in a crisis. *Decisions: Case studies in Australian public policy,* eds. Wilenski, P., Encel, S. & Schaffer, B. Longman, Cheshire, Melbourne.

Chatterton, B. & Chatterton, L. (1986). Conserving native vegetation on private farms in South Australia. *Australia Habitat,* **14.**

Dunsdorfs, E. (1956). *The Australian wheatgrowing industry 1788–1948.* Melbourne University Press, Melbourne.

Fenner, C. (1931). *South Australia – a geographical study, structural, regional and human.* Whitecombe & Tombs, Melbourne & Sydney.

Forster, A. (1866). *Forster's South Australia – its progress and prosperity.* Sampson Low, Son and Marston, Milton House, Ludgate Hill, London.

Gordon, D. J. (1908). In *Handbook of South Australia: progress and resource.* Adelaide.

JAI. (1897). *Journal of Agriculture and Industry,* ed. A. Molineux, Minister of Agriculture, Adelaide.

Light, W. (1984). *William Light's brief journal and Australian diaries.* Introduction and notes by David Elder. Wakefield Press, Adelaide.

Meinig, D. W. (1963). *On the margins of the good earth. The South Australian wheat frontier 1869–1884.* John Murray, London.

Molineux, A. (Ed.) (1882). In *Garden & Field,* **IX.** Under the patronage of the Royal Agricultural and Horticultural Society of South Australia, and the Chamber of Manufacturers.

Molineux, A. (Ed.) (1887). *Garden & Field,* **XII.** Under the patronage of the Royal Agricultural and Horticultural Society of South Australia and the Chamber of Manufacturers.

Perkins, A. J. (1934). Some parting reflections on wheat growing in S.A. *Journal of the Department of Agriculture,* **38.** Adelaide.

Pike, D. (1957). *Paradise of Dissent, South Australia 1829–1857.* Melbourne University Press, Melbourne, and Cambridge University Press, Cambridge.

Price, A. G. (1924). *The foundation and settlement of South Australia, 1829–1845.* Adelaide.

RCI (1904). *Report of Proceedings (1903–4),* **XXXV.** Royal Colonial Institute, London.

Sa'd Abujaber, R. (1989). *Pioneers over Jordan – the frontiers of settlement in Transjordan 1850–1914.* I. B. Taurus, London.

SAYB (1970). *South Australian Year Book,* No. 5. Commonwealth Bureau of Census and Statistics, South Australian office.

Spafford, W. J. (1927). How to improve cereal yields in South Australia. *Journal of Department of Agriculture of South Australia,* **31.**

Springborg, R. (1985). *A critical assessment of the transfer of Australian dryland agricultural technology to the Middle East.* Working Paper 85/2. Australian National University, Canberra.

Thomas, W. K. (1879). *South Australia: an account of its history, progress, resources and present position.* Reprinted from *The S.A. Register,* 6 September. W. K. Thomas & Co, Adelaide.

Williams, M. (1974). *The making of the South Australian landscape*. Academic Press, London.

Chapter 3

Anon. (1987). *La culture intensive du blé*. Institut technique des grandes cultures, Ministère de L'Agriculture et de la Pêche, Alger.
Breakwell, E. J. (1946). Some modern trends in the agriculture of the wheat belt. *Journal of the Department of Agriculture of South Australia*, **50**.
Breakwell, E. J. & Jones, R. H. (1946). Cropping results at Roseworthy College 1945–46. *Journal of the Department of Agriculture of South Australia*, **50**.
Bowden, O. (1940). Wheat growing in the Lower North of South Australia. *Journal of the Department of Agriculture of South Australia*, **44**.
Callaghan, A. R. (1938). Cropping results for 1937 season at Roseworthy Agricultural College. *Journal of the Department of Agriculture of South Australia*, **42**.
Callaghan, A. R. &. Millington, A. J. (1956). *The wheat industry in Australia*. Angus & Robertson, Sydney.
Carter, E. D. (1974). *The potential for increasing cereal and livestock production in Algeria*. Report for CIMMYT, Mexico, and Ministry of Agriculture and Agrarian Reform, Algeria.
Carter, E. D. (1975). *The potential role of integrated cereal–livestock systems from southern Australia in increasing food production in the Near East and North Africa region*. Report for the FAO/UNDP regional project on the improvement and production of field food crops. Waite Institute, Adelaide.
Carter, E. D. (1981). *A review of the existing and potential role of legumes in farming systems of the Near East and North Africa region*. Report to ICARDA in 1978. Waite Institute, Adelaide.
Cook, L. J. (1927). Report on improved pastures. *Journal of the Department of Agriculture of South Australia*, **31**.
Correll, J. (1896). Drilling grain crops with fertiliser. In *Report of Agricultural Bureau of S.A. Eighth Congress*. South Australian Government Printer.
Coulter, R. (1898). Systematic Farming. *Journal of Agriculture and Industry*, **2**.
Dahmane, A. (1987). The role of annual medic species in the improvement of cereal and livestock production in Tunisia – a research review. *Report of the expert consultation on annual medic pasture*. FAO, Rome, and the Bureau of Livestock and Pasture Production, Ministry of Agriculture, Tunisia.
Day, H. R. (1954). The 1953 harvest and current activities at Minnipa Research Centre. *Journal of the Department of Agriculture of South Australia*, **58**.
Donald, C. M. (1982). Innovation in Australian agriculture. In *Agriculture in the Australian Economy*, ed. D. B. Williams, 2nd edition. Sydney University Press.
Dunsdorfs, E. (1956). *The Australian Wheatgrowing Industry 1788–1948*. Melbourne University Press, Melbourne.
Ferguson, A. B. (1935). Improving the pasture. *Journal of the Department of Agriculture of South Australia*, **39**.
French, R. (1963). New facts about fallowing. *Journal of the Department of Agriculture of South Australia*, **67**.
Foster (farmer), (1903). Grass for pasture. *Journal of Agriculture and Industry*, **7**, Ministry of Agriculture, Adelaide.

Hannaford papers. Collection of articles, advertisements and pamphlets held by Alf. Hannaford & Co Ltd, Adelaide. Copies in Chatterton papers, Italy.

Herriot, R. I. (1935). The application of nitrogenous fertiliser to cereal crops. *Journal of the Department of Agriculture of South Australia*, **39**.

Herriot, R. I. (1943). The Cinderella of agriculture. *Journal of the Department of Agriculture of South Australia*, **47**.

Herriot, R. I. (1954). Tillage. *Journal of the Department of Agriculture of South Australia*, **58**.

Higgs, E. D. (*c*. 1981). Can medics make a comeback? (unpublished). Chatterton papers, Italy.

JAI (1897a). Lowrie's reasons for fallow (p. 17) and recommendation for super (p. 711). *Journal of Agriculture and Industry*, **1**. Minister of Agriculture, Adelaide.

JAI (1897b). Lowrie distinguishes between types of fallow. *Journal of Agriculture and Industry*, **1**. Minister of Agriculture, Adelaide.

JAI (1897c). Account of the use of posts, droppers, wire and barbed wire for cheap fences. *Journal of Agriculture and Industry*, **1**. Minister of Agriculture, Adelaide.

JAI (1898). Correll refers to 'native clovers' growing after super has been used on cereals. *Journal of Agriculture and Industry*, **2**. Minister of Agriculture, Adelaide.

JAI (1900). Lehmann (farmer) in *Journal of Agriculture and Industry*, **4**. Minister of Agriculture, Adelaide.

JAI (1901). W. E. Hawke, (farmer from Arthurton) reports on the need to fallow early to control grasses and take-all. *Journal of Agriculture and Industry*, **5**. Minister of Agriculture, Adelaide.

JAI (1902). Accounts of Professor Towar's talks to farmer on dust mulch. *Journal of Agriculture and Industry*, **6**. Minister of Agriculture, Adelaide.

JDA (1909). Angaston Branch meeting report in *Journal of the Department of Agriculture of South Australia*, **7**. (Note: name of Journal changed in 1904/5 from *Journal of Agriculture and Industry*.)

JDA (1910). Wirrabarra branch meeting report in *Journal of the Department of Agriculture of South Australia*, **14**.

JDA (1912). Nantawarra branch report for 9th August 1911 in *Journal of the Department of Agriculture of South Australia*, **15**.

JDA (1927a). Advertisement for Mt Barker sub-clover by an agent for C. Howard in *Journal of the Department of Agriculture of South Australia*, **31**.

JDA (1927b). Pasture improvement competition – first series. *Journal of the Department of Agriculture of South Australia*, **31**.

JDA (1927c). J. Fradd (farmer from Beetaloo Valley) reports to his local Bureau Branch in *Journal of the Department of Agriculture of South Australia*, **31**.

JDA (1939). Callaghan, A. R. & Jones, R. H., 'Results for Roseworthy college farm for the season 1938'. *Journal of the Department of Agriculture of South Australia*, **43**, 298–305.

JDA (1945). J. A. Kelly at a conference discussion at Maitland on Yorke Peninsula. Do the advantages of bare fallow outweigh its disadvantages? *Journal of the Department of Agriculture of South Australia*, **49**.

JDA (1958). Turretfield – field day 1958. *Journal of the Department of Agriculture of South Australia*, **62**.

Lowrie, W. (1898). Roseworthy trials reported in *Journal of Agriculture and*

Industry, **2**. Minister of Agriculture, Adelaide.

McCulloch, R. N. (1956). Roseworthy Agricultural College 1955–56. *Journal of the Department of Agriculture of South Australia*, **60**.

Molineux, A. (ed.) (1886). Experiments at the Agricultural College. *Garden & Field*, **11**. Under the patronage of the Royal Agricultural and Horticultural Society of South Australia and the Chamber of Manufacturers.

Molineux, A. (ed.) (1887). *Garden & Field*, **12**. Under the patronage of the Royal Agricultural and Horticultural Society of South Australia and the Chamber of Manufacturers.

Norton, R. S. & Britza, D. K. (1982–3). Permanent rotation experiment C.l. *Waite Agricultural research institute biennial report*. Adelaide.

Pearson, F. B. (1962). Hay isn't the only way. *Journal of the Department of Agriculture of South Australia*, **66**.

Perkins, A. J. (1911). Fourth report of permanent experiment field, seasons 1909–10 and 1910–11. *Journal of the Department of Agriculture of South Australia*, **15**.

Perkins, A. J. (1918). Rotation of crops. Report of talk given at Kybybolite reported in *Journal of the Department of Agriculture of South Australia*, **12**.

Perkins, A. J. (1934). Some parting reflections on wheat growing. *Journal of the Department of Agriculture of South Australia*, **38**.

Richardson, A. E. V. (1929). The necessity for increased efficiency in wheat production. *Journal of the Department of Agriculture of South Australia*, **33**.

SAYB. (1970) *South Australian Year Book. No. 5*. Commonwealth Bureau of Census and Statistics, South Australian Office.

SAYB. (1985). *South Australian Year Book. No. 20*. Australian Bureau of Statistics, South Australian Office.

Seedco (1988). *Seedco, Seeds of success – a twenty five year history*. South Australian Seedgrowers' Cooperative, Adelaide.

Spafford, W. J. (1918). Turretfield experimental farm harvest report. *Journal of the Department of Agriculture, South Australia*, **22**.

Spurling, M. B. (1987). Agricultural achievements in South Australia. *Journal of the Australian Institute of Agricultural Science*, **53**, (2).

Summers, W. L. (1901). Manuring of grass lands. *Journal of Agriculture and Industry*, **5**. Minister of Agriculture, Adelaide.

Summers, W. L. (1907). Some introduced clovers. *Journal of the Department of Agriculture*, **11**. Minister of Agriculture, Adelaide.

Trumble, H. C. (1938). Barrel medic (*M. tribuloides* Der.) as a pasture legume. *Journal of the Department of Agriculture of South Australia*, **42**.

Webber, G. D., Cocks, P. S. & Jefferies, B. C. (1976). *Farming Systems in South Australia*. South Australian Department of Agriculture and Fisheries, Adelaide (also available in French and Chinese).

Webber, G. D. & Boyce, K. C. (1987). Légumineuse pastorales annuelles dans les rotations céréaliers. *Céréaliculture*, **16**, Institut des techniques des grands cultures, Alger.

Williams, M. (1974). *The Making of the South Australian Landscape*. Academic Press, London.

Williams, S. (1942). Sources of economic loss in wool. *Journal of the Department of Agriculture of South Australia*, **46**.

Wray, J. M. (1934). Report on improved pasture near Kybybolite. *Journal of the Department of Agriculture of South Australia*, **38**.

Chapter 4

Agreement (1974). *Agreement for Pilot Farm*. Council for Agricultural Development, Executive Authority for Jabel el Akhdar.

Anon, (1978). *Harvest in all seasons in Jabel El Akhdar*. Jabel El Akhdar Authority, Libya.

Benkhail, A. S. & Bukechiem, A. A. (1989). Irrigation farming in the Jabel El Akhdar: prospects and problems. *Libya: State & Region – a study of regional evolution*, eds. Allan, J. A., McLachlan. K. S. & Buru, M. M. School of Oriental and African Studies, Centre of Near and Middle East Studies, London University, and Al Fateh University, Tripoli, and The Society for Libyan Studies, London.

Chatterton, B. (1974). Personal diary. Chatterton Papers, Italy.

Chatterton, B. & Chatterton, L. (1982). Notes of Libyan projects. Chatterton Papers, Italy.

Day, H. (1979). Transcribed interview. Chatterton Papers, Italy.

El Akhrass, H., Wardeh, M. F. & Sbetah, A. A. (1988). *Ley farming system and other rainfed systems in the Great Libyan Socialist People's Arab Jamahirya*. ACSAD, Syria.

Kelly, W. J. (1975–7). Personal letters to family from El Marj, Jabel el Akhdar. Copies in Chatterton Papers, Italy.

McPhee, G. R. (1980). *Final report of activities and recommendations of the South Australian Agricultural team based on the demonstration farm at El Marj*. South Australian Department of Agriculture.

Prance, T. M. (1979). Example of a projection based on an economic comparison of the Australian and traditional farming methods, El Marj, Libya. In *Fodders for the Near East: Annual medic pastures*, 97/2. FAO, Rome.

Chapter 5

Allan, J. A. (1989). The effects of the demand for livestock products on natural resources. *Libya: State & Region – a study of regional evolution*, eds. Allan, J. A. McLachlan, K. S. & Buru, M. M. School of Oriental and African Studies, Centre of Near and Middle East Studies, London University, and Al Fateh University, Tripoli, and The Society for Libyan Studies, London, pp. 119–26.

Allen, J. M. (1979). Ley farming in Libya – North Africa. *Wool technology and sheep breeding*, **27** (4), University of New South Wales, Sydney, pp. 5–9.

Chatterton, B. & Chatterton, L. (1987). Increasing livestock production in dry zones – policy options for the Middle East and North Africa. *Land Use Policy*, **4** (2). Butterworth, pp. 121–9.

El Akhrass, H., Wardeh, M. F. & Sbetah, A. A. (1988). *Ley farming system and other rainfed systems in the Great Libyan Socialist People's Arab Jamahirya*. ACSAD, Syria.

Lightfoot, R. J. (1979). *Report on the sheep breeding programme, cereals project, Gefara Plains Authority*. Western Australian Department of Agriculture, Perth.

Seedco. (1988). *Seedco, Seeds of success, a twenty five year history*. South Australian Seedgrowers' Cooperative, Adelaide.

Faint, P. L. & Spencer, P. W. (1979). Unpublished report of Marthuba and Laziette projects. Seedco, Adelaide.

Ianson, M. (1974). Unpublished report on sowing program on 10 Farzhoogha farms. Seedco, Adelaide.
Libyan Harvest Report. (1974). Unpublished. Seedco, Adelaide.
Masters, D. (1979). Unpublished Final report. Seedco, Adelaide.
Reichstein, K. (1979). Unpublished report. Seedco, Adelaide.
Richter, L. K. (1979). Unpublished report on season's contract. Seedco, Adelaide.
Salter, D. L. M. (1976). Unpublished report on shearer instruction course. Seedco, Adelaide.
Schmidt, M. (1979). Unpublished report. Seedco, Adelaide.
Schultz, K. B. (1976). Unpublished report. Seedco, Adelaide.
Anon. (1979). Unpublished report. Seedco, Adelaide.
Treasure, G. L. (1976). Report to Chairman of Jabel el Akhdar Authority on contract, November, 1976. Seedco, Adelaide.
Treasure, G. L. (1982a). Report to Jabel el Akhdar Authority. Seedco, Adelaide.
Treasure, G. L. (1982b). Sidi Hamed demonstration farm El Marj – inspection report March 1982. Seedco, Adelaide.

Chapter 6

Adem, L. (1989). Synthèse de la récherche sur les espèces annuelles de medicago en Algèrie. *XVI Congrès International des Herbages*. Nice.
Anon. (1983). Vers blanc des céréales. *Bulletin phytosanitaire, No. 8*, Avertissements agricoles, Ministère de l'Agriculture et Reforme Agraire, Alger.
Bakhtri, N. (1980). Introduction of medic/wheat rotation in the North African and Near East countries. *Rainfed agriculture in the Near East and North Africa*. FAO, Rome.
Birks, P. (1983). Private communication. Chatterton papers, Italy.
Chatterton, B. (1979). Notes taken during official visit to Algeria. Chatterton Papers, Italy.
Chatterton, B. & Chatterton, L. (1979–90). Notes taken during interviews with Ministers and officials on study tours during which the operation of agricultural policies on farms and research centres was inspected and reports formulated. Chatterton Papers, Italy.
Chatterton, B. & Chatterton, L. (1989). *Fodders for the Near East: annual medic pastures*, 97/2. FAO, Rome.
Chatterton, L. & Marsh, J. (1980). Audio visual representation of South Australian medic farming system. South Australian Department of Agriculture.
DAWPRW (1979). *Dryland agriculture in winter precipitation regions of the world*. Dryland agriculture technical committee, Oregon State University, Cornwallis.
DCSJ (no date). *Développment des Céréales et suppression de la Jachère, 1971-5*. (Note document prepared by the Algerian Ministry of Agriculture and Agricultural Reform and submitted to FAO for the initial planning of the project and the resulting official preparation papers for La Phase 1 du Projet.) Project ALG/71/537. FAO, Rome.
Doolette, J. (1980). Improved crop rotation technology for Tunisia. *Improving Dryland Agriculture in the Middle East and North Africa*, Food Research Institute, Stanford University and the Ford Foundation Middle East Regional Office, Cairo, pp. 72–80.

FAO (1991). *Yearbook, Production volume no. 45.* Rome.

Golusic, A. (1978). Rapport finale. *Développment des céréales et suppression de la jachère, 1971–5.* Project ALG/71/537. FAO, Rome.

ITGC (1987). *Le culture intensive du blé.* Institut technique des grandes cultures. Ministère de l'Agriculture et de la Pêche, Alger.

K.C. Final report (1983). *Ksar–Chellala integrated steppe development project.* 7 volumes. South Australian Department of Agriculture.

Muckle, Y. B. (1978). Développment des céréales et suppression de la jachère, Algérie, machinisme agricole. *Developpment des céréales et suppression de la jachère, 1971–5.* Project ALG/71/537. FAO, Rome.

Pattison, R. J. (1978). End of mission report. *Développment des céréales et suppression de la jachère, 1971–5.* Project ALG/71/537. FAO, Rome.

Pattison, R. J. (1978) *op.cit.* Reference to Table 3.1 (Adem) on variation in percentage protein and total protein of *M. truncatula*, and Table 3.2 (Vercoe & Pearce, 1969) on dry matter digestibility, p. 11.

Preliminary proposals (1976). *Preliminary proposals. Technical and advisory services for pilot project Ksar–Chellala area Algeria 1976.* South Australian Government.

Tebessa Seminar. (1986). Unpublished proceedings of seminaire international sur la strategie d'aménagement et développement de la steppe et des zones arides, 26 April. High Commission for the Steppe, Alger.

Treacher, T. T. (1990). Policy issues in livestock production in arid regions and the management of extensive grazing lands. *Proceedings of FAO expert consultation on strategies for sustainable animal agriculture in developing countries.* FAO, Rome.

Trumble, H. C. (1948). *Blades of Grass.* An Australiana Society publication. Georgian House, Melbourne.

Webber, G. D. (1975). Report on visit to Algeria. *Développment des céréales et suppression de la jachère, 1971–5.* Project ALG/71/537. FAO, Rome.

Webber, G. D., Cocks, P. S. & Jefferies, B. (1976). *Farming Systems in South Australia.* South Australian Department of Agriculture and Fisheries, Adelaide (also available in French and Chinese).

Zeghida, A. (1967). La rotation céréales – medicago dans les zones à vocation céréales-élévage. *Céréaliculture. No. 16.* ITGC. Alger.

Chapter 7

Arer, A (1980). The role of rainfed agriculture in the Near East region: summary of present situation, potential and constraints. *Proceedings of regional seminar on rainfed agriculture in the Near East and North Africa*, Amman, Jordan, May 5–10, 1979. FAO, Rome.

Bull, B. (1984). Notes and observations. *Quarterly report No. 15* (quarter ending 30 September 1984). Jordan dryland farming project. SAGRIC International, Adelaide.

Carter, E. D. (1975). *The potential role of integrated cereal–livestock systems from southern Australia in increasing food production in the Near East and North African region.* Report for the FAO/UNDP regional project on the improvement of production of field food crops. Waite Institute, Adelaide.

Chatterton, B. & Chatterton, L. (1984). Alleviating land degradation and increasing cereal and livestock production in North Africa and the Middle East using annual medicago pasture. *Agriculture, ecosystems and environment,*

2, Elsevier Science Publishers, Amsterdam, pp. 117–29.

Chatterton, B. & Chatterton, L. (1979, 1986, 1987 and 1988). Notes of study tours of Jordan dryland agriculture and medic projects. Chatterton Papers, Italy.

Chatterton, L. (1988). Recipes for rangeland revival. *Ceres*, **21**. FAO, Rome.

Cocks, P. S. (1982). Consultant agronomist's report. Appendix 5 In *Jordon dryland farming project report, 1982 1 (6)*. Overseas Project Division, South Australia Department of Agriculture.

El-Ghonemy, M. R. (1980). Opening address. *Proceedings of regional seminar on rainfed agriculture in the Near East and North Africa*. Amman, Jordan, May 5–10, 1979. FAO, Rome.

El Hurani, Haitham, (1980). Jordanian farmers' perceptions of improved wheat technology. *Improving dryland agriculture in the Middle East and North Africa*, Food research institute, Stanford University, and the Ford Foundation, Middle East Regional Office, Cairo.

El-Nabulsi, A. R. (1980). Project for increasing wheat production in Jordan 1967–78. *Proceedings of regional seminar of rainfed agriculture in the Near East and North Africa*, Amman, Jordan, May 5–10, 1979. FAO, Rome.

El-Sakit, H. (1979). Welcome address. *Proceedings of regional seminar on rainfed agriculture in the Near East and North Africa*, Amman, Jordan, May 5–10, 1979. FAO, Rome.

FAO, (1985). *Report on the FAO. expert consultation on rangeland rehabilitation and development in the Near East. 22–5 October*, Rome.

Gotsch, C. (1980). Introduction and an agenda for research. *Improving dryland agriculture in the Middle East and North Africa*. Food research institute, Stanford University, and the Ford Foundation, Middle East Regional Office, Cairo.

Harvey, D. R. (1984). Field evaluation and financial analysis of Jordan dryland farming project and farmers' demonstration areas in the season 1982–83. Appendix 3. in *Quarterly report No. 15* (quarter ending 30 September 1984) Jordan dryland farming project. SAGRIC International, Adelaide.

Heading, G. (1990). Private communication. Chatterton Papers, Italy.

Hesheiwat, K. & Mahommad, J. (1987). *A study about socio-economic farming systems in Ma'in pilot area*. Jordan Ministry of Agriculture. Forestry Department, Amman.

Hewson, R. & Stensholt, B. (1989). *Agricultural technology in the aid program*. Development Papers No. 2. Australian International Development Assistance Bureau, Canberra.

Heysen, C. (1979). *Feasibility study for the introduction of Australian farming technology to the cereal zone in Jordan*. Overseas Projects Unit, South Australian Department of Agriculture & Fisheries.

ICARDA (1993). *Introducing ley farming to the Mediterranean basin*. eds. Cocks, P., Materon, L., Falcinelli, M. & Christiansen, S. Proceedings of international workshop, Perugia, Italy, 26–30 June 1989. ICARDA, Syria.

JDFP (1981–6). Jordan dryland farming project reports. Phase 1: *Quarterly reports Nos. 1–15* from December 1980 to 30 September 1984; Phase 2: *Quarterly reports Nos. 1–6* from January 1985 to June 1986. SAGRIC International, Adelaide.

Peckover, T. (1984). Farm technicians report, in *Quarterly report No. 15*, (quarter ending 30 September 1984). Jordan dryland farming project, SAGRIC International, Adelaide.

Qureshi, W. (1984). Current trends and possibilities of increasing small ruminant production in the Near East. *Expert consultancy on small ruminant research and development in the Near East*, 23–7 October, 1984, Tunis. FAO, Rome.

Rangeland Management (1984–9). Newsletter containing details of objectives and progress of FAO/UNDP regional rangeland management project RAB/84/025, ed. J. Hall. Tunis.

Reeves, R. (1993). An extension strategy currently used in Jordan to introduce the ley farming system. *Introducing ley farming to the Mediterranean basin*, eds. Cocks, P., Materon, L., Falcinelli, M. & Christiansen, S. Proceedings of an International workshop, Perugia, Italy, 26–30 June 1989, ICARDA, Syria.

Sa'd Abujaber, R. (1989). *Pioneers over Jordan– the frontiers of settlement in Transjordon 1850–1914*. I. B. Taurus, London.

Squires, U. (1981). *Livestock Management in arid zones*. Inkata Press, Melbourne.

Chapter 8

APDP (1982). *Agro-pastoral development project, Erbil, Annual report July 1982*. Overseas project division, Department of Agriculture, Adelaide.

Buringh, P. (1960). Soils and soil conditions in Iraq. In *Final report and recommendations, DDAJP 1985*. Western Australian Overseas development authority, Perth.

Chatterton, B. (1979 and 1980–1). Notes taken during South Australian delegation discussions with Iraqi officials and inspection of sites in Iraq. Chatterton Papers, Italy. Copies in special collection, University of Adelaide.

Chatterton, L. (1983). Notes of South Australian delegation visit to Iraq. Chatterton Papers, Italy.

DDAJP (1985). *Final report and recommendations. Development of dryland agriculture, Jezira project, Northern Iraq, March*. Western Australian overseas projects authority, Perth.

SAGRIC brief (1983). Brief prepared for the Minister of Agriculture's visit to the Middle East and North Africa, January/February, 1983. SAGRIC International, Adelaide.

Special report (1982). Material in brief for Minister of Agriculture's visit to the Middle East and North Africa January/February, 1983. SAGRIC International, Adelaide.

Chapter 9

Carter, E. D. (1974). *The potential for increasing cereal and livestock production in Algeria*. Report for CIMMYT, Mexico, and Ministry of Agriculture and Agrarian Reform, Algeria.

Carter, E. D. (1975). *The potential role of integrated cereal–livestock systems from Southern Australia in increasing food production in the Near East and North African region*. Report for the FAO/UNDP regional project on the improvement and production of field food crops. Waite Institute, Adelaide.

Carter, E. D. (1981). *A review of the existing and potential role of legumes in farming systems of the Near East and North African region*. Report to ICARDA in 1978. Waite Institute, Adelaide..

FAO (1953). *Legumes in agriculture*, eds. Trumble, H. C., Whyte, R. O. & Nilsson-Leissner, G, Rome.

FAO (1956). *Pasture and fodder crops in rotations in Mediterranean agriculture*, ed. P. Oram. Rome.

FAO (1980). *Rainfed agriculture in the Near East and North Africa.* Proceedings of regional seminar held in May 1979 in Amman, Rome.

ICARDA (1993). *Introducing ley farming to the Mediterranean basin*, eds. Cocks, P., Materon, L., Falcinelli, M. & Christiansen, S. Proceedings of International workshop, Perugia, Italy, 26–30 June 1989. ICARDA, Syria.

Trumble, H. C. *Blades of Grass.* An Australiana Society publication. Georgian House, Melbourne.

Chapter 10

ACAPAZ Proceedings. (1987). *Grazing management and annual medicago in pasture development programs for rangeland areas*, Chatterton, B. & Chatterton, L. In Part 2. of *Proceedings of international conference on animal production in arid zones*, Damascus, 7–12 September, 1985. ACSAD, Syria.

Allen, J. M. (1979). Ley farming in Libya – North Africa. *Wool technology and sheep breeding*, **27** (4). University of New South Wales, Sydney, pp. 5–9.

ACSAD (1988). *Ley farming system and other rainfed systems in the Great Libyan Socialist People's Jamahira*, El-Akhrass, H. Wardeh, M. F. & Sbetah, A. A. Syria.

Cocks, P. (1985). *Integration of cereal/livestock production in the farming systems of North Syria.* In proceedings of conference held in Addis Ababa, Ethiopa. ICARDA, Syria.

FAO (1987). *Report of the expert consultation on annual medic pasture in North Africa and the Near East.* Held in Sidi Thebet, Tunisia. Organised by the Bureau of livestock and pasture of the Ministry of Agriculture Tunisia, supported by FAO, Rome.

FAO (1989). *Fodders for the Near East: annual medic pastures.* 97/2. Chatterton, B. & Chatterton, L. FAO, Rome.

FAO (1990). *Development and preservation of low input Mediterranean pastures and fodder systems*, 6th meeting of the FAO. European sub-network on Mediterranean pastures and fodder crops. Held in Bari, Italy. Rome.

Ford Foundation (1980). *Improving dryland agriculture in the Middle East and North Africa.* Food research institute, Stanford University, and the Ford Foundation Middle East Regional office, Cairo.

ICARDA. (1991). *Annual report.* Syria.

ICARDA (1993). *Introducing ley farming to the Mediterranean basin*, eds. Cocks, P., Materon, L., Falcinelli, M. & Christiansen, S. Proceedings of international workshop, Perugia Italy, 26–30 June 1989. ICARDA, Syria.

ICARDA Newsletter (1993). *Dryland pasture & forage legume network news.* Issue No. 8. Syria.

ICARDA. Newsletter (1994). *Dryland pasture and forage legume network news.* Press release 7 March. Syria.

Jaritz, G. & Amine, M. (1989). *Practical experience with the implementation of annual medic-based ley farming system in Morocco.* Summary in English of a report entitled *Evaluation de l'Operation ley farming.* Direction de la production végétale, DCLF/SLF MARA Rabat.

McWilliams, J. R. (1982). *Pasture and forage systems in North Africa and West Asia.* Review of programs and proposals for future research prepared for the Director General, ICARDA, Syria.

Nunn, J. (1981). *Soldier settlers – War service land settlement Kangaroo Island.* Investigator Press, Adelaide.

Nygaard, D. (1980). Farming systems research – Parts 1 and 2. In *Rural Marketing Policy*, Nos. 1 and 2. South Australian Department of Agriculture, Adelaide.

S.A. Government (1980). *Proceedings of International congress on dryland farming.* Vols. 1 & 2, Adelaide.

Springborg, R. (1985). *A critical assessment of the transfer of Australian dryland agricultural technology to the Middle East.* Working paper No. 80/2. Development Studies Centre. Australian National University, Canberra.

Squires, V. & Tow, P. (eds.) (1991). *Dryland Farming: a systems approach. An analysis of dryland agriculture in Australia.* Sydney University Press, and Oxford University Press, Melbourne.

Chapter 11

Chatterton, B. & Chatterton, L. (1986–7). Notes of missions to Tunisia. Chatterton Papers, Italy.

Duthil, J. (1967). *La production fourraggère.* J.-B. Baillière & Fils, Paris.

FAO (1987). *Report of expert consultation on annual medic pasture in North Africa and the Near East.* In Sidi Thebet, Tunisia. Organised by the Bureau of livestock and pasture of the Ministry of Agriculture, Tunisia, supported by FAO, Rome.

Lapeyronie, A. (1982). *Les production fourragères Méditerranéennes.* Maisonneuve, G. P. & Larose, France.

Chapter 12

Adem, L. (1989). Synthèse de la rechèrche sur les espèces annuelles de medicago en Algèrie, *XVI Congrès International des Herbages*, Nice.

Chatterton, B. & Chatterton, L. (1988). *A national plan for medic training, extension and research.* Prepared for Director-General, ITGC, Algeria; ACSAD, Damascus, & IFAD, Rome.

Chatterton, B. & Chatterton, L. (1990). Notes of Algerian mission. Chatterton Papers. Italy.

Chapter 13

Chatterton, B. & Chatterton, L. (1986). Notes of review of Operation ley farming Morocco. Chatterton Papers, Italy.

Chatterton, B. & Chatterton, L. (1989). Notes of review of Abda Plains project in Morocco. Chatterton Papers, Italy.

Jaritz, G. & Amine, M. (1989). *Practical experience with the implementation of annual medic-based ley farming system in Morocco.* Summary in English of a report entitled *Evaluation de l'Operation ley farming.* Direction de la production végétale. DCLF/SLF/MARA Rabat.

Masters, D. (1986). Report of Moroccan contract. Seedco, Adelaide.

Rodda, Q. (1986). Report of Moroccan contract. Seedco, Adelaide.

Chapter 14

ACIAR Annual report (1987–8). Australian Centre for International Agricultural research, Canberra.

Austrade (1987). *Draft export development strategy – Australian agricultural systems*, Australian Department of Trade, Canberra.

Chatterton, B. & Chatterton, L. (1986). The Jackson report and agricultural aid. *Australian overseas aid*, eds. Eldridge, P., Forbes, D. & Porter, D. Croom Helm, Sydney.

Lockeretz, W. (1981). The Lessons of the dust bowl, in *Use and misuse of the earth's surface*, ed. Skinner, B. J. Readings for *American Scientist*, Kaufmann Inc., Los Altos.

Range Plant Handbook (1988). Prepared by United States Department of Agriculture Forest Service and published by the United States Government Printing Office, Washington, D.C. in March 1937; reprinted in a Dover edition by Dover Publications, Inc., 31 East 2nd St., Mineola, N.Y. 11501.

Index